United States Nuclear Regulatory Commission

Protecting People and the Environment

NUREG-1635, Vol.10

Review and Evaluation of the Nuclear Regulatory Commission Safety Research Program

A Report to the U.S. Nuclear Regulatory Commission

Advisory Committee on Reactor Safeguards

AVAILABILITY OF REFERENCE MATERIALS
IN NRC PUBLICATIONS

United States Nuclear Regulatory Commission

Protecting People and the Environment

NUREG-1635, Vol.10

Review and Evaluation of the Nuclear Regulatory Commission Safety Research Program

A Report to the U.S. Nuclear Regulatory Commission

Manuscript Completed: April 2012
Date Published: October 2012

Advisory Committee on Reactor Safeguards

ABSTRACT

This report to the U.S. Nuclear Regulatory Commission (NRC) presents the observations and recommendations of the Advisory Committee on Reactor Safeguards (ACRS) concerning the NRC Safety Research Program being carried out by the Office of Nuclear Regulatory Research. In its evaluation of the NRC research activities, the ACRS considered the programmatic justification for the research, as well as the technical approaches and progress of the work. The evaluation identifies research crucial to the NRC mission. The interdisciplinary effort of the State-of-the-Art Reactor Consequence Analyses Project is not addressed in this report. Early in the project, the ACRS provided reports on the technical approach of this activity.

*****

TABLE OF CONTENTS

FIGURES

TABLES

ABBREVIATIONS

ABWR	advanced boiling-water reactor
ACRS	Advisory Committee on Reactor Safeguards
ADAMS	Agencywide Documents Access & Management System
AECL	Atomic Energy of Canada Limited
AIChE	American Institute of Chemical Engineers
ALARA	as low as is reasonably achievable
AMPs	aging management programs
ANL	Argonne National Laboratory
ANS	American Nuclear Society
AOOs	anticipated operational occurrences
APWR	advanced pressurized-water reactor
ASME	American Society of Mechanical Engineers
ASP	Accident Sequence Precursor Program
ASTM	American Society for Testing and Materials
ATLAS	Advanced Thermal-Hydraulic Test Loop for Accident Simulation
ATWS	anticipated transient without scram
BFBT	boiling-water reactor full-size fine-mesh bundle tests
BIP	Behavior of Iodine Project
BWR	boiling-water reactor
CAROLFIRE	cable response to live fire
CCF	common-cause failure
CFD	computational fluid dynamics
CFR	*Code of Federal Regulations*
CHRISTI-FIRE	cable heat release, ignition, and spread in tray installations during fire
CODAP	Component Operational Experience, Degradation and Aging Program
CONOPS	concept of operations
CRDM	control rod drive mechanism
CRPPH	Committee on Radiation Protection and Public Health
CSARP	Cooperative Severe Accident Research Program
CSAU	code scaling, applicability, and uncertainty
DESIREE-FIRE	direct current electrical shorting in response to exposure fire
DI&C	digital instrumentation and control
DNB	departure from nucleate boiling
DOE	U.S. Department of Energy
DVI	direct vessel injection
EAC	environmentally assisted cracking
EDMGs	extensive damage management guidelines
EIS	environmental impact statement
EMUG	European MELCOR user group
EOPs	emergency operating procedures
EPA	U.S. Environmental Protection Agency
EP Act	Energy Policy Act of 1992
EPIX	Equipment Performance and Information Exchange System
EPMDA	expanded proactive materials degradation assessment
EPR	evolutionary power reactor
EPRI	Electric Power Research Institute
ESBWR	economic simplified boiling-water reactor
ESP	early site permit

ABBREVIATIONS (Cont'd)

FCI	fuel-coolant interaction
FFF	fuel fabrication facility
FMEA	failure modes and effects analysis
FSME	Office of Federal and State Materials and Environmental Management Programs
GDC	general design criteria
GEH	General Electric-Hitachi
GSI	Generic safety issue
GUI	graphical user interface
GWd/t	gigawatt day per metric ton
HAMMLAB	Halden Man-Machine Laboratory
HDPE	high-density polyethylene
HERA	human event repository and analyses
HF	human factor
HFE	human factors engineering
HRA	human reliability analysis
HRP	Halden Reactor Project
HSI	human-system interface
HTGR	high-temperature gas-cooled reactor
I&C	instrumentation and control
IAEA	International Atomic Energy Agency
IASCC	irradiation-assisted stress-corrosion cracking
ICRP	International Commission on Radiological Protection
ICSP	international collaborative standard problem
IEEE	Institute of Electrical and Electronics Engineers
IFRAM	International Forum for Aging Management
IGSCC	intergranular stress-corrosion cracking
INPO	Institute of Nuclear Power Operations
IPEEE	individual plant examination of external events
IPWRs	integral pressurized-water reactors
ISG-TP	International Steam Generator Tube Integrity Program
ISI	inservice inspection
ISP-50	International Standard Problem 50
JAEA	Japan Atomic Energy Agency
JCCRER	Joint Coordinating Committee for Radiation Effects Research
JNES	Japan Nuclear Energy Safety Organization
KAERI	Korea Atomic Energy Research Institute
KM	knowledge management
LANL	Los Alamos National Laboratory
LER	licensee event report
LERF	large early release frequency
LOCA	loss-of-coolant accident
LP/SD	low -power/shutdown
LRGDs	license renewal guidance documents
LSTF	large-scale test facility
LWR	light-water reactor

ABBREVIATIONS (Cont'd)

MACCS	MELCOR Accident Consequence Code System
MASLWR	multiapplication small light-water reactor
MCAP	MELCOR Cooperative Assessment Program
MCCI	molten core concrete interaction
MCNP	Monte Carlo N-Particle Transport Code System
MDEP	Multinational Design Evaluation Program
MeV	million electron volts
MORs	monthly operating reports
MOX	mixed oxide
MSPI	Mitigating System Performance Index
NAS	National Academy of Science
NASA	National Aeronautics and Space Administration
NCRP	National Council on Radiation Protection and Measurements
NDE	nondestructive examination
NEA	Nuclear Energy Agency
NEI	Nuclear Energy Institute
NFPA	National Fire Protection Association
NGNP	next generation nuclear plant
NMSS	Office of Nuclear Material Safety and Safeguards
NOR	nitration oxidation reaction
NPPs	nuclear power plants
NRC	U.S. Nuclear Regulatory Commission
NRO	Office of New Reactors
NRR	Office of Nuclear Reactor Regulation
NSIR	Office of Nuclear Security and Incident Response
NTTF	Near-Term Task Force
NUPEC	Nuclear Power Engineering Corporation
OECD	Organization for Economic Cooperation and Development
ORNL	Oak Ridge National Laboratory
OSU	Oregon State University
PARCS	Purdue Advanced Reactor Core Simulator
PCI	pellet cladding interaction
PCMI	pellet cladding mechanical interaction
PCT	peak clad temperature
PINC	Program for the Inspection of Nickel Alloy Components
PIRT	phenomena identification and ranking table
PNNL	Pacific Northwest National Laboratory
PRA	probabilistic risk assessment
PSBT	Pressurized-water reactor subchannel and bundle test
PSHA	probabilistic seismic hazard analysis
PSI	Paul Scherrer Institute
PSU	Pennsylvania State University
PUMA	Purdue University Multidimensional Integral Test Assembly
PWR	pressurized-water reactor
PWSCC	primary water stress corrosion cracking
QRSM	Quantitative Software Reliability Model
R&D	research and development
RBHT	Rod Bundle Heat Transfer

ABBREVIATIONS (Cont'd)

RCS	reactor coolant system
RES	Office of Nuclear Regulatory Research
RG	regulatory guide
RPV	reactor pressure vessel
RTF	radioiodine test facility
SAMGs	severe accident management guidelines
SARNET	Europe's Severe Accident Research NETwork
SCALE	standardized computer analysis for licensing evaluation
SCC	stress corrosion cracking
SERENA	steam explosion resolution for nuclear applications
SDP	significance determination process
SGTR	steam generator tube rupture
SMR	small modular reactor
SNAP	Symbolic Nuclear Analysis Package
SNL	Sandia National Laboratories
SOARCA	state-of-the-art reactor consequence analyses
SPAR	standardized plant analysis risk
SPAR-AHZ	standardized plant analysis risk all hazards models
SRM	staff requirements memorandum
SSCs	structures, systems, and components
SSE	safe shutdown earthquake
SSHAC	Senior Seismic Hazard Analysis Committee
SSWICS	small-scale water ingression and crust strength
TMI-2	Three Mile Island Unit 2
TRACE	TRAC-RELAP advanced computational engine
U.S.	United States
USGS	United States Geological Survey
V&V	verification & validation
VHTR	very-high-temperature reactor
xLPR	extremely low probability of rupture

1. INTRODUCTION

In this report, the Advisory Committee on Reactor Safeguards (ACRS) presents the results of its review and evaluation of the U.S. Nuclear Regulatory Commission (NRC) Safety Research Program. The NRC maintains a safety research program to:

- Ensure its regulations and regulatory processes have sound technical bases and these bases are refined as new knowledge develops.

- Prepare for anticipated changes in the nuclear industry that could have safety implications.

- Develop improved methods to carry out its regulatory responsibilities.

- Maintain an infrastructure of expertise, facilities, analytical capabilities, and data to support regulatory decisions.

The current research program, organized by the Office of Nuclear Regulatory Research (RES), is closely coupled to specific, near-term issues to support regulatory activities and initiatives in the Offices of Nuclear Reactor Regulation (NRR), New Reactors (NRO), Nuclear Material Safety and Safeguards (NMSS), Nuclear Security and Incident Response (NSIR), and Federal and State Materials and Environmental Management Programs (FSME). RES has succeeded over the last few years in its effort to tie research activities it undertakes to near-term issues confronted by the NRC line organizations (NRO, NRR, NMSS, NSIR, and FSME).

For the purposes of this report, the ongoing research has been examined in terms of the following technical disciplines:

- advanced reactor designs

- digital instrumentation and control systems

- fire safety

- reactor fuel

- human reliability and human factors

- materials and metallurgy

- neutronics and criticality safety

- operational experience

- probabilistic risk assessment

- radiation protection

- nuclear materials and waste

- seismic and structural engineering

- severe accidents and source term

- thermal hydraulics

The interdisciplinary research effort of the state-of-the-art reactor consequence analyses (SOARCA) project is not addressed in this report. Early in the project, the ACRS provided reports on the technical approach of this activity.

Chapter 2 of this report provides a synoptic account of research activities in each of the technical disciplines and highlights some of the accomplishments of the work. Additional details on each of the research areas are included in chapters 4 through 17.

In its review of the NRC Safety Research Program, the ACRS has focused on the technical and regulatory justification for the ongoing research activities. The ACRS supports research that achieves the following:

- Identifies and resolves current safety and regulatory issues.

- Provides technical bases for the resolution of foreseeable safety issues.

- Develops the capabilities of the agency to independently review risk-significant proposals and submittals by licensees and applicants.

- Supports agency initiatives, including the move toward a much greater use of risk information in the regulatory process and to evolve NRC safety regulations to be "technology neutral."

- Improves the effectiveness and efficiency of the regulatory process.

- Maintains technical expertise within the agency and associated facilities in disciplines crucial to the agency mission that are not readily available from other sources.

The accident at Fukushima Dai-ichi in Japan has increased interest in accident initiators, accident management, and consequences for accidents that are within the design basis, as well as those beyond the design basis. Research to help determine the implications of the accident at Fukushima for the U.S. nuclear reactor fleet is discussed in Chapter 3. The NRC line organizations will benefit from research initiatives that strengthen the technical understanding of this accident.

2. GENERAL OBSERVATIONS AND RECOMMENDATIONS

In this chapter, the ACRS highlights major components of the ongoing research activities dealing with the safety of nuclear power plants and presents its recommendations concerning these activities.

General Observations

The NRC has succeeded over the last few years in its effort to tie research activities it undertakes to near-term issues being confronted by the NRC line organizations (NRO, NRR, NMSS, NSIR, and FSME). Two-thirds of research activities support specific needs of these offices. One-third are mandated programs such as the Accident Sequence Precursor (ASP) program. A small portion of the research budget focuses on long-term research subjects expected to be critical in 5 to 10 years.

The strategy for the identification of research subjects, through "user need" documents, together with the associated process for the prioritization of research needs, has worked reasonably well. Research activities are yielding useful products to the line organizations in a timely manner. One area of concern is that research focused on line-organization needs may be terminated prematurely precluding in-depth understanding that would be of use in the future.

Major research activities often include collaborations with other Federal agencies, industrial institutions, and international partners. Such collaborations provide timely and thoughtful peer input while giving the agency the ability to leverage its expertise and resources on key topics of common interest. In addition, such collaborations provide an opportunity to help train new NRC staff members as they participate in these multiparty research efforts.

Future Research Focus Areas

In its 2010 report to the Commission on the NRC Safety Research Program (NUREG-1635, Vol. 9, "Review and Evaluation of the Nuclear Regulatory Commission Safety Research Program: A Report to the U.S. Nuclear Regulatory Commission"), the ACRS noted the growing emphasis on the use of numerical simulation to resolve reactor safety issues. In particular, it was noted that a recent U.S. Department of Energy (DOE) initiative, with the participation of both licensees and nuclear steam suppliers, is underway to establish the so-called high fidelity simulation of the thermal hydraulics, neutronics, and mechanics of nuclear power plants. While this initiative is a long-term project, it is indicative of the DOE and industry intent to use computational tools, such as computational fluid dynamics (CFD), in thermal hydraulics research to address safety issues. The ACRS continues to believe that the NRC needs to be ready to properly evaluate products of such numerical simulations.

The ACRS acknowledges the growth in the use of computations to resolve reactor safety issues. However, we remain concerned that this has not been accompanied by appropriate experimental testing capable of yielding high resolution and high fidelity data suitable for validating computational predictions.

Experimental data are needed to provide an adequate technical basis for many safety decisions. When using best-estimate realistic analyses considering uncertainty, such data should be used to validate analyses that rely primarily on numerical simulations. Examples of such safety decisions are modifications to plant design or power uprate requests from the plant operator. In these cases, the original operating envelope can be reassessed based on a rigorous approach to determine the operational margins and plant performance. The NRC first demonstrated this process with the code scaling, applicability, and uncertainty (CSAU) methodology. The CSAU methodology provides necessary transparency for both the technical and regulatory communities, while providing focus to the research and development that is necessary to achieve closure of specific regulatory issues. It provides for a series of process steps grouped into categories, many of which have the role of establishing credibility of the analyses, as well as identifying appropriate test validation. Such an approach requires fidelity in geometric, physics, and material models, code verification, and most importantly, experimental validation and uncertainty quantification.

Experimental validation provides evidence that models are accurate for a hierarchy of subsystem and full system tests relevant to the intended application. Multidimensional and multiphysics numerical simulations must be validated through comparison of code predictions to experimental data. This is best accomplished in a hierarchal manner that includes separate effects tests, subsystem tests, and full system tests, when practical. In addition, uncertainty quantification provides the evidence necessary to make risk-informed decisions by quantifying variabilities and uncertainties in key decision metrics and in understanding which input uncertainties are most important.

Unfortunately, experimental facilities within the United States, available to the NRC for obtaining experimental validation, continue to dwindle. There is hope that adequate experimental facilities remain in the rest of the world and that these facilities are available to the NRC through formal as well as informal collaborations. The NRC has demonstrated leadership in matching needs with international facilities wherever they can be obtained. The NRC should continue to develop proactive strategies that ensure access to such facilities. This can be accomplished through either partnering with others to develop new test capabilities or taking advantage of foreign capabilities in the areas of research requiring data (e.g., natural hazards, coupled neutronics thermal hydraulics, materials, severe accidents, and risk analysis). In addition, to maintain expertise and to train new hires in experimentation, the NRC should consider assignments for promising younger staff at international experimental facilities.

Major Observations on Individual Areas of Research

Major observations, conclusions, and recommendations concerning specific research activities are summarized below. Additional details on the research activities in the various technical disciplines are provided in chapters 4 through 17

Advanced Reactor Designs

The high-temperature gas-cooled reactor (HTGR) research plan is focused on the next-generation nuclear plant (NGNP), and the final very-high temperature reactor (VHTR) design has yet to be identified. DOE funding for the NGNP program is decreasing. Without an applicant or a specific design, existing research must be examined and refocused.

Safety research for small modular reactor (SMR) designs is just beginning. Staff has identified some important issues for licensing of these designs. These efforts provide a significant input to the development of safety research necessary for future reactor licensing activities.

Digital Instrumentation and Control Systems

The ACRS is pleased to see a comprehensive research plan for digital instrumentation and control (DI&C). This plan was developed based on consultation with NRC program offices.

A new critical element being added to DI&C research is cyber security. Its major feature is to ensure security against external malicious threats by using one-way-only communication from digital safety systems to the outside world. This data barrier effort was developed in response to a user need identified during the licensing review process for a DI&C safety system upgrade to an existing operating plant.

For a new reactor plant design application, the staff maintains that cyber design review during licensing is prohibited by Title 10 of the *Code of Federal Regulations* (10 CFR) 73.54, "Protection of Digital Computer and Communications Systems and Networks." This interpretation is reflected in Regulatory Guide (RG) 1.152, "Criteria for Use of Computers in Safety Systems of Nuclear Power Plants," which explicitly excludes the review of DI&C design features to provide satisfactory cyber security during the licensing process.

The ACRS continues to see the lack of integration of safety and cyber security in the design stage as an impediment to achieving quality, secure DI&C safety systems. The NRC recommends an integrated approach during the DI&C licensing process.

Fire Safety

The NRC's fire safety research program is at the forefront of a rapidly evolving understanding of fire events and the assessment of risks from fire damage in nuclear plants.

The focus and priorities for current research projects are determined primarily by user-identified needs. This process is responsive to immediate and near-term technical issues, and it should remain an important part of integrated planning. However, longer term research is needed to address emerging issues and longer term requirements that are not fully dictated by current user needs.

New reactor designs rely heavily on integrated digital instrumentation, protection, and control systems. Extensions of operating licenses and obsolescence of analog equipment also likely will lead to replacement of many currently installed instrumentation and control (I&C) systems with digital platforms. Limited information is available on the effects from fire, heat, and smoke on digital equipment. Research projects should address issues such as the effects from fire damage and heat on fiber optic cables, the effects of heat on digital equipment, and smoke damage to digital signal processing and computation modules.

Reactor Fuel

As nuclear fuel is designed and operated to higher burnup levels, new fuel claddings are being developed to replace the Zircaloy and Zirlo^{TM} alloys explicitly cited in current regulations. NRC research has done an excellent job developing a sound technical basis for regulatory qualification of future zirconium-based fuel claddings.

NRC regulations provide criteria and guidance for the prevention of large numbers of fuel cladding failures caused by large and rapid power increases during anticipated operational occurrences (AOOs). Failure mechanisms addressed by the regulations include departure from nucleate boiling (DNB), mechanical strain caused by pellet cladding mechanical interaction (PCMI), and fuel center melting. However, the pellet clad interaction (PCI) stress corrosion failure mechanism, which is equally capable of causing large numbers of fuel failures during AOOs, is not addressed in the regulations. Research should be pursued to correct this deficiency. There is a wealth of experimental data and operational experience available to the NRC that can be used to develop and validate an empirical

PCI failure model for use in regulatory decisionmaking. This development should be focused on the prevention of PCI fuel failures of current and future fuel designs during AOO power excursions.

Human Factors and Human Reliability

The NRC human factor (HF) and human reliability analysis (HRA) research program has evolved into a carefully coordinated series of projects that are extending knowledge in this area and providing useful products. For example, RES took the initiative to extend the work of the International HRA Empirical Study to a U.S. HRA Empirical Study, including experiments conducted at a U.S. nuclear power plant simulator using that plant's operators and procedures.

Materials and Metallurgy

The ACRS supported the initial vision of the Proactive Materials Degradation Assessment Program and has recommended its continuation. In our previous research report, we commented that the program seemed to have lost its momentum and focus. Target items of low knowledge and high-to-medium susceptibility to degradation had been identified. There was an expectation that the next step should be the validation of one or more of these predictions by experimental means or by focused inspections at operating plants. However, in light of the additional operating experience highlighted through the license renewal process, the current approach of revisiting the assessment of materials degradation and expanding the scope to include concrete and cable insulation seems warranted and will provide a firmer basis for experimental programs or new nondestructive examination (NDE) or monitoring programs.

The ACRS is supportive of RES's active participation in international efforts relating to materials degradation, such as the International Forum for Aging Management (IFRAM) Research. Such international efforts were pioneered by the NRC over 30 years ago and remain valuable and cost effective today.

Neutronics and Criticality Safety

The current NRC research programs on neutronics and criticality safety are well supported by user needs and by the agency obligations under the Energy Policy Act of 2005, and they appear to be progressing well.

Even greater demands on neutronics and criticality analyses can be anticipated in the future. Much of this demand arises because licensees of existing and planned reactors are going to greater lengths to optimize core designs. Optimization often involves reductions in the symmetry of the core and concomitant increases in complexity contributing to the need for larger scale computations. Furthermore, there is greater demand for uncertainty and sensitivity analyses of these computational results. Together, these evolutions are leading to more intensive uses of computational capabilities, and they create the need for faster code execution. The NRC is sponsoring efforts to develop parallel computational capabilities in the standardized computer analysis for licensing evaluation (SCALE) code suite.

Operational Experience

The Operational Experience Research program is being managed and executed in a manner consistent with the needs of the NRC. It provides data and tools necessary for regulatory decisionmaking and for the assessment of regulatory effectiveness. There appears to be good coordination between the research staff and the user organizations.

The events at the Fukushima Dai-ichi nuclear plant site highlight the need to strengthen and integrate the various onsite emergency response capabilities. RES should ensure that its research program provides the necessary technical basis needed for enhancement and integration of the various onsite emergency response capabilities and the development of integrated command and control and decisionmaking structures.

Probabilistic Risk Assessment

A substantial fraction of probabilistic risk assessment (PRA) research resources is allocated to routine maintenance and updating of standardized plant analysis risk (SPAR) models and supporting data. It is not evident that maintenance of these functions should be an RES obligation. RES should focus on needed advancements of state-of-the-art PRA capabilities.

Comprehensive PRAs for all new reactors and risk-informed applications for an increasing number of upgraded operating plants must include coherent models for complex digital instrumentation, protection, and control systems. The ACRS continues to emphasize the need to understand and define a common set of integrated hardware and software failure modes for these systems. The ACRS recommends that failure mode research efforts should have top priority, and that clear project goals should be set.

Methods to systematically identify, document, and quantify sources of uncertainty are not applied consistently throughout the agency. The ACRS recommends that RES initiate efforts to ensure that an appropriate characterization of uncertainty is performed in all agency analyses.

Radiation Protection

The staff has developed an appropriate and robust research program in the areas of radiation protection. This program includes radiation protection of workers and radiological assessments related to radiation exposure and health risks to the general population around NRC-licensed nuclear facilities.

Nuclear Materials and Waste

The NRC continues to maintain a robust program of research related to nuclear materials and radioactive waste topics related to licensing, facility siting, facility environmental performance, and the decontamination and decommissioning of licensed facilities.

Seismic and Structural Engineering

The NRC seismic and structural research program in support of regulatory activities has been exemplary. There is a well-developed research plan that has been broadly reviewed for both technical quality and programmatic impact.

Severe Accidents and Source Term

The ACRS supports the strategy that the NRC staff has developed to support regulatory decisions for severe accidents via computer code development validated by experimental data analysis and evaluation. This approach has successfully allowed the NRC to maintain and update its modeling capabilities for severe accident analyses. Planned program extensions and continuations of these collaborations are well worth the investment.

MELCOR assessments of the Fukushima Dai-ichi accident have identified some deficiencies in boiling-water reactor (BWR) specific modeling capabilities. The ACRS recommends that the NRC expand its current severe accident research program to obtain the required data to enhance and validate models that are found to be deficient.

Thermal Hydraulics

RES has moved forward with a small but effective effort on CFD analysis of safety-related problems involving important multidimensional phenomena. This effort should continue and should include the integration of such CFD analyses with the essentially one-dimensional analyses provided by system codes such as TRAC-RELAP advanced computational engine (TRACE).

Advanced light-water reactor systems will employ natural circulation cooling for normal and off-normal operation. These designs should not be certified based on analysis alone. Confirmatory experimental data should be obtained. International collaboration is one approach to obtaining this information. It takes advantage of facilities that are of a scale and capability that currently do not exist in the United States. Furthermore, this approach allows the NRC to draw on the expertise of international partners, who have continued to maintain a very high level of capability in the thermal-hydraulics field. However, the ACRS continues to recommend that complementary development of national facilities to address safety-related thermal-hydraulics issues be seriously considered.

3. SAFETY RESEARCH IN THE AFTERMATH OF EVENTS IN FUKUSHIMA DAI-ICHI NUCLEAR COMPLEX IN JAPAN

In its review and evaluation of the NRC Safety Research Program, the ACRS has identified several topical areas for research that would support NRC line organizations in implementing lessons learned from Fukushima to the U.S. nuclear reactor fleet. The ACRS encourages RES to develop an integrated plan to obtain the necessary technical basis for implementing the lessons learned with respect to:

- protection from external hazards

- protection from severe accidents

- emergency response and severe accident management capabilities

- accident-tolerant instrumentation to characterize plant response

- improved understanding of severe accident phenomena

Each of these topical areas is discussed in this section.

Protection from External Hazards

Fukushima demonstrated that the combined effects of a major seismic event and subsequent massive tsunami can disable the majority of structures, systems, and components (SSCs) required to ensure core cooling and fission product containment. Similarly, tornadoes, floods and hurricanes acting singly or in combination may cause extended loss of offsite power with coincident physical damage to nonsafety structures or equipment at multiple units. Damage from severe storms or other site-specific hazards also may disable external essential cooling water that provides cooling from the ultimate heat sink. Vulnerabilities to those hazards and their attendant damage may not be completely addressed from assessments that focus only on bounding deterministic extreme natural phenomena (e.g., design-basis seismic or flooding events).

The ACRS encourages RES to develop a risk-informed approach for evaluating the protection against extreme natural phenomena, whether acting singly or in combination. This approach is not currently required by regulation, but it can provide a context for a rational approach to a defense-in-depth philosophy for protecting against such extreme events.

A case in point would be the evaluation of the potential effects from a tsunami on a nuclear power plant near coastal areas. The NRC does not currently require a probabilistic approach to tsunami hazard evaluation for operating plants. However, a probabilistic approach to such a hazard evaluation is becoming more prevalent in many hazard analyses (e.g., for seismic events in Regulatory Guide (RG) 1.208, "A Performance-Based Approach to Define the Site-Specific Earthquake Ground Motion").

The advantage of this approach is its use in determining the likelihood of an event and including the full range of uncertainty in input parameters. Deterministic approaches to tsunami hazard have been the general rule in current hazard evaluations and have proven to be useful to coastal engineers in developing effective tsunami counter-measures. Nevertheless, a probabilistic tsunami hazard analysis would serve to elucidate the likelihood of a range of tsunami conditions at a specific plant site and define the range of uncertainties of the parameters involved in tsunami modeling at a site. Development and application of such an approach for tsunamis or floods would be a real advancement in a full scope PRA.

Protection against Severe Accidents

For severe accidents, historically, only a few direct regulatory requirements, such as emergency planning, were instituted. Severe accident regulatory decisions have mostly dealt with reducing the likelihood of such a serious accident rather than coping with one. This approach was based on the assumption that, because of the "defense-in-depth" design philosophy, such accidents are of sufficiently low probability that mitigation of their consequences is not necessary for public safety. The 1979 accident at Three Mile Island Unit 2 (TMI-2), led to the reexamination of the design basis and the consideration of regulations for protection against severe accidents.

The events at the Fukushima Dai-ichi Nuclear Power Station have once again provided an impetus for the re-assessment of regulatory decisions for protection against severe accidents. RES can play an important role in providing the technical basis for any such reevaluation. Hydrogen control can be considered as a specific example. The first significant regulatory action for severe accident mitigation was the hydrogen rule (10 CFR 50.44, "Combustible Gas Control for Nuclear Power Reactors") issued soon after the TMI-2 accident. When this rule was promulgated, it was thought that inerting the containment atmosphere in BWRs with Mark I and Mark II containments was sufficient to eliminate any concerns about combustible gas control during a degraded core accident for these plant designs.

The Fukushima accident has shown this logic is not always valid. Several of the units at Fukushima Dai-ichi experienced explosions in their reactor buildings causing substantial damage, hampering recovery efforts by the operators. Currently, the NRC staff is gathering more details on how the hydrogen may have been released into the reactor building before recommending specific combustible gas control measures in BWRs with Mark I and Mark II containments. Possible paths for hydrogen release into the reactor buildings at the Fukushima plants include degradation of drywell head seals, leakage through damaged valves and couplings in vent lines, and damage to downcomer bellows venting into the suppression pools. Nevertheless, the possible mechanisms for hydrogen release to the surrounding reactor building raise concerns about this issue for all BWRs with Mark I and Mark II containments. Clearly, the issue is not merely a peculiarity of the Fukushima reactors. Informed actions for combustible gas transport, mixing, and control in the surrounding reactor buildings will need input from staff research efforts.

The Fukushima accident demonstrates that hydrogen combustion events can cause significant structural damage. Such damage also can cause debris to fall into the spent fuel pools with subsequent potential ramifications not considered previously. The ACRS has suggested as a defense-in-depth measure that for BWR plants with Mark I and Mark II containments, combustible gas control measures be implemented in the reactor buildings. RES should consider if a research project is needed to support the decisions in which measures are most appropriate (e.g., igniters, passive hydrogen recombiners, or hydrogen getters) and the strategy for their spatial arrangement. Additionally, research focused on identifying and minimizing potential hydrogen leakage paths in aging BWRs (e.g., degradation of drywell head seals and lines venting into the suppression pools) would be highly beneficial, especially if life extension beyond 60 years is contemplated.

Emergency Response and Severe Accident Management Capabilities

The ACRS encourages RES to take an active role in international research efforts to understand the factors affecting the range of operator performance during and following the Fukushima Dai-ichi accident. The accident took operators into complex conditions, with multiple competing demands, and it progressed well beyond the design basis into conditions that would have required exercising severe accident management guidelines (SAMGs) in the United States.

There is little experience in this regime in the nuclear industry, and current training drills do not force operators into such situations. Research into the issues driving human response under such conditions is essential in the long-term response to recommendations to integrate emergency operating procedures (EOPs), the SAMGs, and the extensive damage mitigation guidelines (EDMGs). An international research effort that includes Japanese experts and operators might be effective in revealing elements of operator response that are culturally driven from more universal cognitive complexities.

The Near-Term Task Force (NTTF) report and corresponding staff recommendations expressed the need to strengthen and integrate the various onsite emergency response capabilities (i.e., EOPs, SAMGs, and EDMGs). Such integration could focus on the need to clarify the transition points, command and control, decisionmaking, and training requirements.

The ACRS also has recommended that the onsite emergency response capabilities be expanded to include the plant fire response procedures. ACRS has noted that enhancement and integration of the various onsite emergency response capabilities and the development of command and control and decisionmaking structures (as identified in NTTF recommendations) would be a complex effort requiring substantial interactions among licensees' operations, engineering, and management personnel, industry owners groups, vendor experts, and regulators. Use and integration of the guidance also will require extensive testing. Although the ACRS views these efforts as a long-term endeavor, the NRC recommends that work begin immediately.

The ACRS anticipates that the nuclear industry may develop accident management advisory tools. It is important that the NRC be in a position to evaluate the technical accuracy and utility of such tools. These tools require accurate plant data for key parameters to model an accident as it progresses beyond the design basis with a potential range of operator actions. Such a tool can be used to better inform the possible range of operator actions that can be employed for accident management, to assist in the development of a consistent set of SAMGs, or to audit what has been developed by the licensees for effectiveness and completeness.

Accident Tolerant Instrumentation to Characterize Plant Response for Risk-Dominant Scenarios

The experience at Fukushima showed that essential reactor and containment instrumentation should be enhanced to better withstand beyond-design-basis accident conditions. Immediately after the tsunami flooded the Fukushima Dai-ichi plant, key instrumentation for the reactor vessel, drywell, and wetwell were unavailable for Units 1 and 2 because of loss of alternating current (ac) and dc (direct current) power sources. The instruments at Unit 3 lost power nearly 30 hours later. When power was restored, the validity of data from available sensors was questionable. Pressure transients and seawater addition adversely affected water level accuracy. Thermocouples attached on exterior vessel surfaces were exposed to temperatures above their operating range and were no longer reliable. Evaluations also indicate that some pressure gauges gave erroneous readings.

Robust and diverse instrumentation that can better withstand severe accident conditions is needed to diagnose, select, and implement accident mitigation strategies and monitor their effectiveness. Similar observations were made on the adequacy and availability of instrumentation after the

1979 TMI-2 event. In the 1990s, research sponsored by the NRC demonstrated that it is possible to address this issue by implementing a systematic methodology that includes tasks to identify the required plant information, the location and operating range of sensors currently available to provide such information, and the environmental conditions that such sensors must withstand during risk-dominant accident sequences.

The NRC has recognized the need for enhanced reactor and containment instrumentation and is in the process of adding this to the implementation of the NTTF recommendations. RES should evaluate if updated research is needed to evaluate the adequacy of proposed reactor and containment instrumentation enhancements.

Improved Understanding of the Severe Accident Phenomena

RES should proactively engage in efforts to define and participate in programs that capture and analyze data from the Fukushima incident. The ACRS notes that RES is already working with the DOE and industry to attempt to 'reconstruct' the Fukushima accident to gain greater detailed insights into BWR severe accident phenomena.

The detailed information needed to understand the events and progression of accidents at Fukushima will parallel the information collected in the aftermath of the accident at Three Mile Island. The collection effort at Three Mile Island took substantial time. It was inefficient and conflicted with efforts to recover and stabilize the damaged reactor. Similar conflicts can be anticipated in any effort to collect post-accident data and materials from the damaged Fukushima reactors. A well-developed, broadly accepted and endorsed plan to collect this information will be essential.

The NRC recommends that RES continue its existing efforts with DOE and participate in the development of an international program to obtain and evaluate material samples from these plants to gain insights about accident progression phenomena and physical processes that occurred during the Fukushima accident. Such sample gathering and associated evaluation activities can enhance the understanding and modeling of severe accident phenomena, including BWR melt progression, radionuclide release and transport, the effects of untreated raw water addition, the effects of hydrogen transport and combustion, the operability of safety systems, and ex-vessel phenomena.

4. ADVANCED REACTOR DESIGNS

Background

In the past few years, NRC research on advanced reactor designs has focused on assessment of its research infrastructure needs and the agency's planned safety research to support its review of DOE's NGNP. DOE has interacted with reactor designers, potential process heat users, and industrial and international organizations to support the NGNP design development needs for the thermal-spectrum gas-cooled graphite-moderated reactor concept, the so-called VHTR. The DOE effort was reinforced by the passage of the Energy Policy Act of 2005 (EPAct)[1], which authorized appropriation of funds for research and construction activities for the NGNP project. DOE selected the VHTR as the lead design concept for the NGNP. Specifically, the VHTR program is focused on this type of reactor design, and the EPAct authorizes the NRC to collaborate with DOE in safety research related to licensing issues as the project proceeds through licensing to construction and operation.

Recent interest in small modular reactors (SMRs), with indications that applications might be received for one or more SMRs, has generated the corresponding need to identify safety research needs to support review of licensing applications for these designs. The SMRs currently identified as most likely to submit applications in the near term are the integral pressurized-water reactors (iPWRs) such as NuScale and mPower.

Artist's rendition of an iPWR plant

The NRC has completed initial activities to provide the proper background information for the lead designs. Such activities included:

- Development of key information sources for the technologies.

- Conduct of a phenomena-identification and ranking table (PIRT) process for each design to identify the key safety phenomena that require additional research and development. These key phenomena are associated with tools, standards, data, etc., required for design and licensing review of such reactor technology.

Research conducted for advanced reactor designs is closely coordinated with severe accident and source term research and with research on human factors, digital instrumentation and control systems, and materials and metallurgy.

Current Research Activities

For the VHTR, the staff is focused on the development of appropriate evaluation models, methods and guidance using information from past prismatic gas-cooled reactor designs and pebble-bed designs, as well as the current conceptual VHTR designs from the industrial teams working

---

[1] See Subtitle C: Next Generation Nuclear Plant Project.

with DOE and its national laboratory contractors.

Based on the 2001 HTGR NRC Research Plan, research topics the staff is considering include:

- plant safety analysis, including thermal fluids and accident analysis

- nuclear analysis

- fuel performance and fission product behavior

- high-temperature materials performance

- graphite performance

- safety issues related to process heat applications

- structural analysis (with particular focus on high-temperature effects)

The research plan is well written and comprehensive. However, because the final VHTR design has yet to be determined, the plan is focused on the research and development (R&D) aspects that are applicable to the range of possible VHTR reactor designs. In addition to the topics listed above, the plan covers several other technical disciplines in which ongoing or planned generic R&D for light-water reactors will be applicable to HTGRs with appropriate modifications: instrumentation and control for high-temperature applications, human factors and human reliability analysis, probabilistic risk assessment, and risk-informed infrastructure development.

One area that has received much attention is the development of evaluation models and tools covering four areas: thermal-fluid analysis, nuclear analysis, fuel performance, and fission product release and transport.

The staff has identified a number of data needs for the HTGR. Issues that specifically deserve attention include steam-graphite oxidation, models for graphite dust production during operation and necessary limitations on dust accumulation, and approaches for dealing with uncertainty in the selection of mechanistic scenarios.

Work on SMRs is in its early stages, but the staff has identified some safety issues that are likely to be important for licensing decisions. Technical issues related to scaling, integration, design basis, and multimodule human-systems interface are especially important with respect to safety and licensing.

Assessment and Recommendations

The 2001 HTGR NRC Research Plan is heavily focused on NGNP, and the final VHTR design has yet to be identified. Recent congressional authorizations suggest that DOE funding for this program is decreasing. Without an applicant or a specific design, existing research must be examined and refocused.

Safety research efforts on SMR designs is in its early stages. The NRC has identified important issues for licensing of these designs. These efforts provide significant input into the development of safety research necessary for future reactor licensing activities.

5. DIGITAL INSTRUMENTATION AND CONTROL SYSTEMS

As nuclear power plants transition to digital, software-based technology for reactor protection and engineered safeguards systems, it is crucial to ensure that the design application of this technology preserves the critical attributes of redundancy, independence, deterministic processing behavior, and defense in depth and diversity necessary to ensure the reliable shutdown and actuation of protection and safeguards systems. In addition, cyber security must be added to the list of critical attributes.

Safety systems involve simple actuations and do not involve complex feedback and control functions. While software-based digital systems offer the designer much greater flexibility and functionality than analog systems, they bring a much increased potential for loss of safety division independence through interdivision data communication and new modes of common cause failure in software programming.

With the above attributes in mind, the 2010—2014 Research Plan has been divided into five major topic areas:

- safety aspects of digital systems

- security aspects of digital systems

- advanced nuclear power concepts

- knowledge management

- carryover projects from the fiscal year (FY) 2005–2009 research plan

Within these areas are 27 projects, seven of which carry over from the FY 2005–2009 plan. These projects were developed based on consultation with NRC program offices to serve their needs.

Candidate Generation III Nuclear Power Plant Control Room

One of the key stumbling blocks in assessing safety assurance of software-based systems is the lack of agreement among industry and staff on effective methods for failure (fault) modes and effects analysis (FMEA) for regulatory assurance for complex logic in DI&C systems.

There are several projects underway to develop a better understanding of which frameworks may be viable for this purpose, including expert elicitation studies, failure mode investigations, and a project to assess the basic utility of FMEA for these systems from a fundamental perspective.

Another major challenge is portraying the reliability of software-based systems in PRAs. Several projects have resulted in the issuing of NUREGs that benchmarked methods for reliability modeling and assessing traditional PRA methods for DI&C systems. Another project is being initiated to develop models of digital safety systems integrated with reactor safety analysis tools, as well as for use in PRAs.

One of the key safety aspects in establishing safety assurance in DI&C systems is well-designed and controlled software. Within the knowledge management focus area, six draft RGs (for existing RGs that are almost 15 years old) are being updated to more current Institute of Electrical and Electronics Engineers (IEEE) standards. These draft RGs deal with critical software attributes such as verification and validation (V&V), configuration management, test documentation, software unit testing, software requirements specifications, and life cycle management for safety systems.

The new critical attribute being added to the mix of concerns is cyber security. Several initiatives have been completed to assess the security vulnerabilities of digital computing platforms. The results of these assessments have been provided to NRC staff, as well to the platform vendors. The issue becomes how to get licensees to update their software to mitigate the vulnerabilities identified in these assessments. A path needs to be identified to resolve this issue since current policy and regulatory guidance do not have enforcement authority. Compliance is all voluntary.

In addition, a major feature to ensure security from external malicious threats is the incorporation of one-way-only communication from digital safety systems to the outside world. RES sponsored a cyber assessment of a hardware device used as a data barrier between security levels 4 and 3 systems (basically from a safety to nonsafety level). It was reverse engineered by Sandia National Laboratories (SNL), and the final evaluation deemed it satisfactory for its application.

The evaluation noted above was conducted during the licensing review process for a DI&C safety system upgrade to an existing operating plant. However, for a new reactor plant design application, the review of safety system design via RG 1.152 explicitly prohibits the review of DI&C design features for their ability to provide satisfactory cyber security design during the licensing process. The cyber security features according to RG 5.71, "Cyber Security Programs for

Nuclear Facilities," will not be evaluated until after the design is complete (upwards of 3 or 4 years after licensing) and that review will be performed by NRC site inspectors that may not be experienced in design of digital systems and in cyber security threats and solutions. The ACRS identified this issue and, at this point, the staff has maintained that any additional actions are prohibited by 10 CFR 73.54. Thus, there is not an integration of design and cyber security during the DI&C licensing process. The NRC continues to see this nonintegrated approach as a vulnerability to achieving quality, secure DI&C safety systems.

Another project of interest is the evaluation of the use of wireless technology and networks interfacing with DI&C safety systems. While studies have been completed to investigate the implications of using wireless technology, no specific application for the use of wireless technology has been presented to the ACRS at this point. While gaining an understanding of the technology and its vulnerabilities is paramount, the acceptance of its use in interfacing safety systems should be viewed with skepticism. Modeling its reliability and its resistance to electromagnetic and radio frequency interference is problematic at best and its cyber security risks are enormous. Intuitively, providing guidance for its use in the traditional regulatory guidance framework may not be rigorous enough and may require more prescriptive approaches, which will challenge the regulatory execution policies presently in place.

The recent events at Fukushima illuminated the need for improved, hardened instrumentation that can survive anticipated severe accident environments and provide critical plant-condition information. Examples include reactor pressure, temperature, and water level conditions inside the reactor vessel; spent fuel pool temperature and actual water level, real-time hydrogen concentrations at key locations (not a sampling system), and the location of core materials. Such data are critical for the operators to correctly assess the plant status and implement appropriate severe accident mitigation strategies. Determining if core cooling was being achieved was hampered by the lack of reliable instrumentation. In addition, this new instrumentation needs to be either nonelectrical or have dedicated power sources that are available for weeks and possibly longer without recharging. Consideration of these needs should be evaluated for future research projects.

6. FIRE SAFETY

Background

The current focus of the fire safety research program is to support the NRC's regulatory needs during licensee transitions to risk-informed, performance-based fire protection programs that meet the requirements of 10 CFR 50.48(c) and the referenced 2001 edition of the National Fire Protection Association (NFPA) Standard NFPA-805, "Performance-Based Standard for Fire Protection for Light Water Reactor Electric Generating Plants.". Continuing research activities also develop improved information to support deterministic fire protection programs for licensees that do not adopt a risk-informed approach. In addition to direct support for currently operating reactors, the fire safety research programs also provide input to new reactor licensing reviews and assessment of the risk from fires in new reactor designs.

Current Research Activities

The NRC's research activities in fire safety are depicted in Figure 1. Research activities primarily support programs in NRR and, to a lesser extent, NMSS. These activities are grouped into four technical areas:

- fire modeling
- fire testing
- fire and electrical systems analysis
- fire risk assessment

There are also efforts underway on fire research knowledge management. The following sections briefly discuss the major projects in each area.

DESIREE-FIRE
The objective of the Direct Current Electrical Shorting in Response to Exposure Fire (DESIREE-FIRE) testing program, performed under collaborative research agreement with EPRI, is to evaluate the response of dc circuits to fire conditions. The results of this research project will be used to develop more realistic models of cable fires for use in fire PRA analyses.

Fire Modeling

Fire models provide a phenomenological basis for the evaluation of fire growth, detection, and suppression, and the analysis of potentially risk-significant fire scenarios.

In 2007, the NRC and the Electrical Power Research Institute (EPRI) completed a collaborative project for V&V of five commonly used fire modeling codes (NUREG-1824, "Verification and Validation of Selected Fire Models for Nuclear Power Plant Applications").

That effort identified strengths and weaknesses in each code with respect to specific fire modeling issues. In 2008, a PIRT exercise was completed to assess the predictive capabilities of these fire models for a number of postulated fire scenarios (NUREG/CR-6978, "A Phenomena Identification and Ranking Table (PIRT)

Exercise for Nuclear Power Plant Fire Modeling Applications"). A collaborative project between RES and EPRI uses the results of the V&V of fire models, along with the PIRT evaluations, to develop an integrated "Nuclear Power Plant Fire Modeling Application Guide," NUREG-1934, for the approved models. The guide will describe the capabilities of each code to evaluate specific phenomena during realistic fire scenarios, including limitations, precautions, and lessons learned from practical applications. The guide will benefit risk analysts who use the fire models to evaluate potentially important fire scenarios, and it will support inspection efforts that use the models as input to the significance determination process (SDP).

Future research projects are planned to extend the current fire modeling capabilities to address specific limitations documented in the V&V reports and the PIRT evaluations and to improve predictions of phenomena observed during fire experiments.

Fire Testing

This element of the research program includes two major projects.

The Cable Heat Release, Ignition, and Spread in Tray Installations during Fire (CHRISTI-FIRE) program is performing full-scale fire tests to measure the heat release rates and flame spread characteristics from fires in bundles of electrical cables. The tests are conducted with realistic cable tray configurations and loadings, and they include cables with thermoplastic and thermoset insulations. The quantitative data collected by this project will be used to develop more realistic cable fire models for PRAs and to enhance the predictive capabilities of fire modeling codes.

Small-scale fire tests are being performed to evaluate the performance of spent nuclear fuel shipping cask seals during beyond-design-basis fires that exceed the manufacturers' rated temperatures. This information will be used by NMSS to further develop risk insights related to the transportation of spent fuel.

Fire & Electrical Systems Analysis

The experience from actual fire events confirms that damage to electrical cables may disable equipment and cause unexpected responses from I&C signals. Realistic evaluation of fire-induced circuit damage, particularly involving spurious signals caused by "hot shorts," is a very significant effort in fire risk assessments and in pilot applications of the methods described in NFPA-805 and NUREG/CR-6850, "EPRI/NRC-RES Fire PRA Methodology for Nuclear Power Facilities".

The Direct Current Electrical Shorting In Response to Exposure-Fire (DESIREE-FIRE) project will extend the CAROLFIRE test program (NUREG/CR-6931, "Cable Response to Live Fire (CAROLFIRE)," to examine fire damage to dc circuits. The tests are being performed in collaboration with EPRI. They will include realistic cable tray configurations and loading, and they will evaluate the effects from fire damage to circuits for dc-operated components and typical I&C applications.

An electrical circuit PIRT is currently underway that is examining the results from industry-sponsored cable tests, CAROLFIRE, and DESIREE-FIRE. Upon completion of the electrical circuit PIRT, a structured PRA expert elicitation process will examine the available data and develop conditional probabilities and uncertainties for various ac and dc circuit failure modes that may be caused by fire-induced cable damage.

Fire Risk Assessment

Integration of fire risk into a full-scope PRA framework is a very important research activity. The requirements of NFPA-805 and the guidance in NUREG/CR-6850 provide cornerstones for the development of risk-informed, performance-based fire protection programs and a comprehensive assessment of fire risk.

A collaborative project with EPRI is evaluating elements of HRA methods that apply to post-fire mitigation actions. The goal of this project is to recommend specific methods and guidance for the modeling and quantification of human errors during fire scenarios. FINAL NUREG-1921, "EPRI/NRC-RES Fire Human Reliability Analysis Guidelines," is expected to be published in 2012.

In 2008, NMSS sponsored a quantitative risk assessment to evaluate the risk from potential fires and explosions caused by nitration oxidation reaction (NOR) at the mixed-oxide (MOX) fuel fabrication facility (FFF). This effort included a pioneering attempt to use the methods of PRA to identify the process of highest concern and probability of a NOR event. The feasibility of using probabilistic methods rather than the usual integrated safety assessment methods for nuclear facilities is noteworthy. The results from this probabilistic study were used to support the NMSS safety evaluation for licensing the MOX FFF.

Pilot applications of the methods described in NFPA-805 and NUREG/CR-6850 have been completed. A supplement to NUREG/CR-6850 was issued in September 2010 and incorporates lessons learned from the NFPA-805 Frequently Asked Questions program and improvements to fire modeling and fire-induced damage assessments from ongoing research activities.

Fire Research Knowledge Management

Fire research continues to be a rapidly evolving element of the RES mission. Compilation and dissemination of the information gained from this research, including new results and insights, is vital for understanding the current state of knowledge and planned near-term advancements. Four excellent resources prepared by RES include:

- "The Browns Ferry Nuclear Plant Fire of 1975 and the History of NRC Fire Regulations," NUREG/BR-0361, January 2009. This DVD preserves the history of the Browns Ferry fire and documents its influence on the development of enhanced fire protection regulations.

- "Fire Protection and Fire Research Knowledge Management Digest," NUREG/BR-0465, Revision 1, February 2009. This CD contains a compilation of fire-related reference materials that are useful for NRC inspectors, reviewers, licensees, and other stakeholders. It includes 10 CFR Part 50, "Domestic Licensing of Production and Utilization Facilities," guidelines for fire protection, fire inspection manuals and procedures, generic letters, bulletins, information notices, RGs, and fire-related NUREG reports.

- "A Short History of Fire Safety Research Sponsored by the U.S. Nuclear Regulatory Commission, 1975–2008," NUREG/BR-0364, June 2009. This report provides a historical perspective on NRC-sponsored fire safety research, summaries of current research activities, and planned near-term research programs.

- "Methods for Applying Risk Analysis to Fire Scenarios (MARIAFIRES) 2008," NUREG/CP-194 July 2010. This report captures the 2008 NRC/EPRI joint training classes and provides a self study program that can be viewed at the user's convenience.

These references are updated periodically.

Assessment and Recommendations

The NRC's fire safety research program is at the forefront of a rapidly evolving understanding of fire events and the assessment of risks from fire damage in nuclear power plants. Structured collaboration with industry provides cost-effective solutions and technical insights that surpass independent efforts. The NRC is a leader in national and international fire safety research.

The focus and priorities for current research projects are determined primarily by user-identified needs. This process is responsive to immediate and near-term technical issues, and it should remain an important part of integrated planning. However, research priorities and programs should more actively anticipate emerging applications and intermediate- to long-term requirements that are not fully dictated by current user needs.

Data and insights from the CAROLFIRE tests have considerably advanced the understanding of fire-induced cable damage. It is expected that the results from current projects such as CHRISTI-FIRE and DESIREE-FIRE will also provide valuable knowledge. The NRC should continue to encourage and support additional testing and fire experiments through collaborative efforts with U.S. industry and international organizations.

Other current research projects also are evaluating HRA methodologies. The goal of those activities is to recommend or develop an integrated methodology that captures the best practices and reduces the variability of assessments from the current assortment of HRA methods. The resulting integrated HRA methodology should apply equally to scenarios that are initiated by internal events, internal hazards (e.g., fires and floods), and external events (e.g., earthquakes) during any plant operating mode (e.g., full power, low power, or shutdown). The fire research described above is focused primarily on only one facet of this problem (i.e., human performance after fires that occur during full power operation). This work should be more closely coordinated with other RES activities to avoid potential duplication or divergence of NRC-sponsored HRA research and conclusions.

Draft NUREG/CR-7114 "Methodology for Low Power/Shutdown Fire PRA," was published in December 2011. This guidance has been developed for the assessment of fire risk during low power and shutdown operating modes. Many of the current fire research programs and results are directly applicable to the evaluation of fire damage during these operating conditions. However, additional information is needed to evaluate such issues as fire initiation frequencies, human-caused fires, effectiveness of detection and suppression, and propagation of heat and smoke through compromised fire barriers that are uniquely associated with personnel activities and SSC configurations during plant shutdown.

New reactor designs rely heavily on integrated digital instrumentation, protection, and control systems. Extensions of operating licenses and obsolescence of analog equipment also will likely lead to replacement of many currently installed I&C systems with digital platforms. Limited information is available on the effects from fire, heat, and smoke on digital equipment. Research projects should address such issues as the effects from fire damage and heat on fiber optic cables, the effects of heat on digital equipment, and smoke damage to digital signal processing and computation modules.

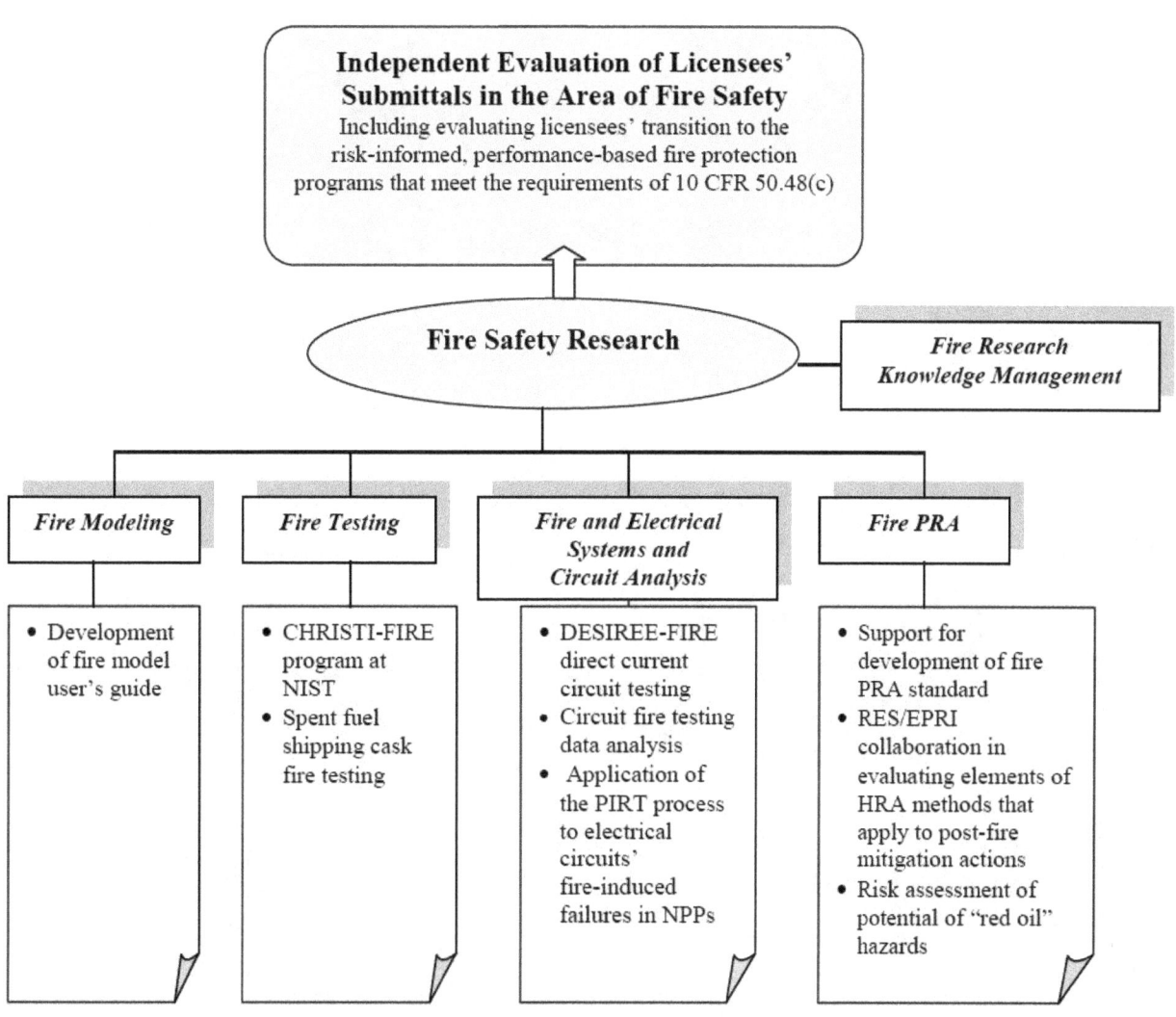

Figure 1. Current NRC Research Activities in Fire Safety

7. REACTOR FUEL

Background

Fuel integrity is an important contributor to nuclear safety. Fuel failures that may occur during normal operation or AOOs are not safety significant events. However, large numbers of fuel failures can occur during a single AOO event, and should be prevented. With the exception of the PCI mechanism, current regulations adequately address the mechanisms that can cause such failures. Extensive efforts have been made to operate U.S. reactors with zero fuel failures. Currently, over 90 percent of U.S. reactors are operating without fuel failures. The various mechanisms capable of causing fuel failures during normal operation have been identified and effectively addressed by design, materials, and operational improvements.

However, power uprates and increased fuel burnup will result in more fuel operating near peak power levels and under conditions that generate more fission products within the fuel rods and greater radiation damage, corrosion, and hydriding of the fuel cladding. Manufacturers are addressing these more demanding requirements with fuel rods and assembly design changes and with the introduction of new fuel pellet, cladding, and assembly structural materials.

The fuel industry is now developing more innovative designs. Lead use assemblies are in operation in which chemical additives are being used to improve fission gas retention and mechanical properties of the UO_2 fuel pellets. Similarly, new fuel cladding materials are in the development and regulatory pipeline. These materials changes address the need to improve the corrosion and hydriding resistance of fuel cladding during normal operation, as well as

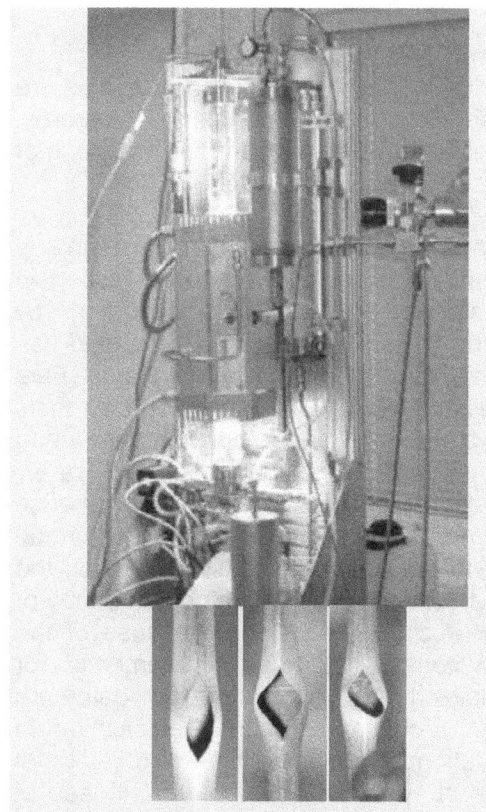

LOCA apparatus operating at 1,200 degrees Celsius at the Studsvik Nuclear AB laboratory (Sweden), where integral tests are being conducted on irradiated, high-burnup fuel rods. Rupture openings reveal the absence of fuel in the ballooned region caused by relocation and dispersal.

its resistance to embrittlement during loss of coolant accidents.

The focus of NRC-sponsored research should remain on events in which large numbers of fuel elements can fail. However, in view of fuel development trends, the NRC also must maintain an adequate research program to ensure its capability to evaluate and license future fuel designs. Active participation in collaborative domestic and international research programs addressing the performance of new designs under normal conditions and AOOs should continue.

Current Research Activities

Current NRC fuel behavior research is depicted in Figure 2. Research activities are focused on various experimental programs, as well as development and maintenance of analytical capabilities.

The suitability of new cladding materials is determined by extensive vendor testing in laboratories and test reactors and by performance evaluations of lead-use assemblies in operating power plants. Fuel suppliers typically perform laboratory tests to evaluate cladding embrittlement during LOCAs. However, fuel suppliers rarely validate predictions of fuel performance during DBA conditions with experimental data. In the past, the industry relied on the NRC to provide information on fuel behavior under accident conditions. Because of the cost, complexity, and requirements for specialized test, much of the data are currently obtained through participation in international collaborative experimental research programs, such as those at Studsvik or Halden.

Experimental data on the performance of fuel and cladding systems under severe accident conditions are not provided by licensees. Severe accidents, of course, pose the bulk of the risk to public health and safety. Experimental data obtained in research sponsored by the NRC show that cladding interactions with fuel play an important role in the rate and extent of core degradation and fission product release and fuel dispersal under severe accident conditions. For example, four recent single rod LOCA experiments performed on irradiated, high-burnup fuel rods in the Studsvik hot cells (see graphic) have demonstrated conditions under which very high burnup fuel can fragment, relocate into the ballooned volume, and subsequently be ejected from the cladding. These results are consistent with earlier LOCA experiments conducted at the Halden reactor project. In addition, research is yielding valuable information on the strength of the ballooned region and the ductility of the balance of the cladding away from the ballooned region. These data will be invaluable in determining future high burnup regulatory limits.

The NRC is in the process of completing experimental studies of high-burnup fuel and cladding behavior under DBA conditions. The NRC has sponsored inpile tests of fuel behavior during reactivity insertion events and out-of-pile tests of fuel behavior under design-basis LOCA conditions. These investigations have been conducted using an impressive combination of national and international collaborations. Results of the research have led to proposals for changes in the regulations. The proposed changes would make the regulations more realistic and could decrease the burden on both the staff and licensees, especially as new fuel and cladding systems are proposed. Implementation of these proposed changes to the regulations has been slow. Acceptance of the changes to the regulations now awaits results of additional testing that could be regarded as confirmatory in nature.

RES is also sponsoring an experimental research program at the Oak Ridge National Laboratory (ORNL) to test high-burnup fuel rods under several dry cask conditions and to determine the influence of hydride reorientation on cladding response to crush loads, mechanical stiffness, and bending. The results of these tests will be used to develop further guidance in meeting the regulatory requirements for storage casks and transportation of spent fuels.

Under an educational grant from the NRC, Pennsylvania State University (PSU) is investigating the fracture toughness behavior for through-thickness crack growth using thin sheet specimens of cold-worked and stress-relieved Zircaloy-4 as a model material. The primary objective of this study is to identify the micro-mechanisms that control crack growth resistance as a function of hydride microstructure and temperature.

The staff also has completed revisions of its fuel performance computer codes, FRAPCON and FRAPTRAN. These computer codes are used to independently confirm analyses done by the vendors and other licensees. The modifications allow the computer codes to be used to evaluate fuels taken to burnups of up to 62 gigawatts day per metric ton (GWd/t) and to evaluate the performance of mixed oxide fuels. MOX fuels have been irradiated in the Catawba reactor as part of a DOE program to dispose of excess weapons-grade plutonium. The NRC also is examining the severe accident behavior of high-burnup and MOX fuels in its severe accident research program.

Assessment and Recommendations

The upgraded fuel performance models appear to meet most near-term agency needs. There is ample capability to assess risks of fuel failure caused by fuel center melting or departure from nucleate boiling events. However, the NRC still does not have empirical or analytical capability to quantitatively assess the more likely risk of PCI fuel failures during AOOs for current or future fuel designs. This fuel failure mechanism may compromise fuel integrity as licensees continue to uprate power, increase fuel burnup, and introduce new fuel and cladding materials. The NRC is not developing a quantitative capability to assess fuel vulnerability to PCI during operational transients. There is a wealth of experimental data on the PCI susceptibility of fuel cladding available to the NRC to develop and validate a practical PCI failure model for use in regulatory decisionmaking without the need for additional testing.

Advances in computing power and computational simulation are making it possible to examine fuel performance in vastly more detail than has been done with either FRAPCON or FRAPTRAN. Whether such detail is needed will depend critically on what efforts are made by licensees to extend fuel burnups beyond the current regulatory limit of 62 GWd/t and the amount of experimental data provided to support these proposed changes to regulatory limits.

There is widespread expectation that the nuclear industry may propose extension of burnup limits to 85 GWd/t. Certainly, a recent research strategy document prepared for DOE and the nuclear energy industry proposes that fuel burnups be extended to 85 GWd/t. There appears to be some confidence within the nuclear industry that such extensions of fuel burnup can be done by extrapolating the currently available bases of fuel performance data and models. Emergence of new physics complicated such extrapolations of fuel performance for burnups beyond 40 GWd/t.

This necessitated the experimental research on fuel behavior under accident conditions that the NRC is currently completing. It is not evident that no new phenomena will arise in connection with the extrapolation of fuel burnup to 85 GWd/t. Consequently, there will be a continuing need for the agency to independently evaluate the safety of proposed changes in the nature and burnup limits of reactor fuels.

Lead-test assemblies of MOX fuels have emerged from their second cycle of irradiation in the Catawba reactor. These MOX fuels are being tested as part of a DOE program to dispose of excess weapons-grade plutonium by using it as reactor fuel. It appears that the NRC does not have plans for any research examinations of these novel fuels either after the first cycle of irradiation or after subsequent cycles of irradiation. In light of the limited NRC experience with MOX fuels, the ACRS recommends that there be a research program to closely follow the post-irradiation examination of the lead-test assemblies planned by DOE.

The NRC must maintain expertise in the area of reactor fuel to respond to future design changes. Because of license renewal, power uprates and the prospect of new reactors, it is anticipated that the vendors will introduce new fuel and cladding systems. The challenge in maintaining what amounts to an essential core competency of the agency arises because of the limited availability of expertise outside of the agency that is independent of licensees. There is a need to groom new talent in the field.

A major difficulty in doing so is the decline in the United States of in-pile test facilities and hot cells for examinations of irradiated fuels and cladding. Long-term collaboration with international partners, having the facilities and staff capable of undertaking pertinent in-reactor studies, may be essential for the NRC to maintain an adequate level of expertise in reactor fuels. One option available to the NRC for the development of fuel staff expertise is the stationing of promising staff members at appropriate international projects, such as the Halden project, for 2 to 3 years. This opportunity has been widely used by industry and should be considered by the NRC.

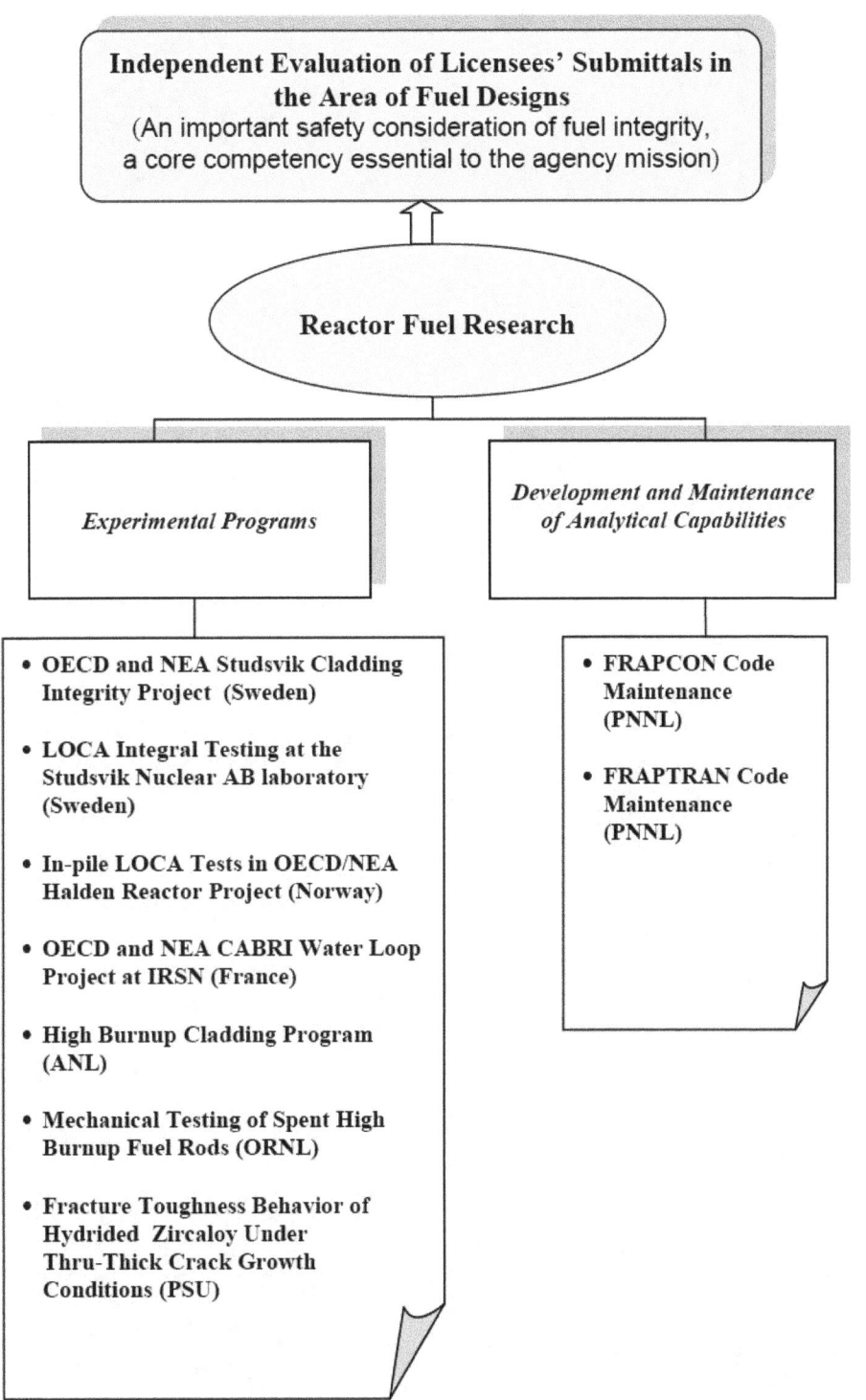

Figure 2. Current NRC Research Activities in Reactor Fuel

8. HUMAN FACTORS AND HUMAN RELIABILITY

Human performance remains central to the safe operation of nuclear facilities. The staff's research activities are focused on improving our understanding of the issues affecting human performance and developing tools for analyzing performance, as well as developing systems to improve the operating environment to enhance the opportunities for improving performance.

Research funding is primarily associated with user needs requests from NRR, NRO, and NMSS. To be successful in meeting long-term research needs, managers work with counterparts in those organizations to craft projects that meet both current and future needs. To achieve an appropriate balance, the staff has organized its activities around the following goals:

- maintaining the infrastructure of expertise, facilities, capabilities, and data

- ensuring that HF and HRA methods and guidance have sound, up-to-date technical bases

- improving HF and HRA methods to reduce uncertainty and promote the state of the art

- expanding the HF and HRA infrastructure for new applications (anticipated changes in industry)

The staff has developed an integrated HF and HRA research program that balances its support of these four goals (see Figure 3). The staff's participation in domestic and international collaborations strengthens these activities and should be encouraged. These collaborations include the Organization for Economic Cooperation and Development (OECD), EPRI, National Aeronautics and Space Administration (NASA), and professional organizations,

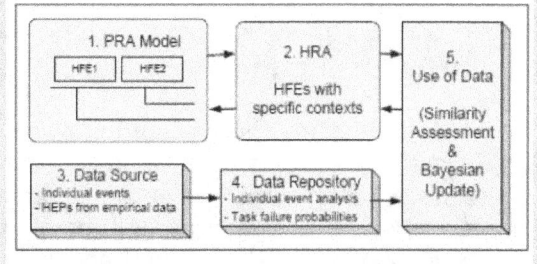

The concept of using empirical data to inform the human error probabilities of human failure events

such as American Nuclear Society (ANS), IEEE, and American Institute of Chemical Engineers (AIChE). The staff has arranged for international experts to be involved in many of its research efforts.

The staff has identified a number of high-priority research needs, and two are in the HF and HRA areas. The NRC has taken a leading role in the International HRA Empirical Study, whose objective is to benchmark HRA model predictions against operating crew performance data developed at the Halden Reactor Project. The results of the first two phases of this study were published in 2009 and 2010, with a third experiment run at a U.S. nuclear power plant in 2011. Several HRA methods are evaluated in this study, strengths and weaknesses are identified, and insights are derived for improving various facets of these methods. For example, method capability to identify and evaluate human performance under various situations, and the extent to which methods are applied as intended by the developers, are addressed in this study.

The adequacy of data available for HRA is a concern for the credibility and consistency of human error probability estimates. To address this need, the staff is developing the Human Event Repository and Analysis (HERA) system, which will support both

HRA and human factors. The objective is to make empirical and experimental human performance data from commercial nuclear power plants available in a format suitable to support HRA.

Human Factors Research

New NPPs will differ from the current plants in several ways that will change how operators interact with the plant: many employ passive designs, all will employ extensive use of DI&C systems, and new human-system interfaces (HSI) will be used. The staff convened a group of experts from research organizations, vendors, and utilities to prioritize research needs. NUREG/CR-6947, "Human Factors Considerations with Respect to Emerging Technology in Nuclear Power Plants," provided the results of this process, which led to a number of followon research projects addressing possible negative impacts of advanced DI&C systems, the concept of operations (CONOPS) for new designs, and the technical basis for staffing verification for advanced control room designs. We commend the staff for taking such a systematic approach to the prioritization and development of research topics.

Other programs are examining whether new computerized procedures can provide adequate safety, developing measurement tools to evaluate workload, situational awareness, and teamwork, developing qualitative operator models to predict operator performance and errors, and developing a technical basis for regulatory decisionmaking about safety culture. All these projects have the potential to improve operator performance and provide improved bases for regulation.

Human Reliability Analysis Research

Since the earliest days of PRA, researchers and practitioners have sought methods for the analysis of human performance during accident scenarios that offer face validity and clearly demonstrable accuracy. The result has been a proliferation of methods, none completely satisfying to all interested parties.

The need for resolution has been recognized by the Commission in a November 8, 2006 SRM directing the ACRS to "work with the staff and external stakeholders to evaluate the different Human Reliability models in an effort to propose either a single model for the agency to use or guidance on which model(s) should to be used in specific circumstances." In response, the staff initiated a series of research projects. The International HRA Empirical Study used the Halden Man-Machine Laboratory (HAMMLAB) to simulate accident sequences. It is a multinational, multiteam effort co-sponsored by the OECD Halden Reactor Project, the Swiss Federal Nuclear Safety Inspectorate, and EPRI. The objective is to collect and analyze crew performance data, to separately apply HRA models to predict crew performance, and to evaluate these models on the basis of a comparison of the simulator data with the model predictions. Following a pilot study and a followon study, the staff organized a U.S. HRA Empirical Study, again using the expertise of HAMMLAB personnel, in a series of experiments run at a U.S. nuclear power plant.

The results of the three sets of experiments have been examined by an international group that developed a number of summary conclusions and a strong recommendation that all HRA analysts develop first a solid and thoroughly documented qualitative analysis to guide later quantification. The NRC staff has used these conclusions and recommendations as the starting point for developing a hybrid HRA method, borrowing the best aspects of other methods applied during the three empirical studies. The goal is to have a single method, or a limited set of alternative methods with a common qualitative underpinning, to apply to specific HRA problems.

Because of fixed schedule demands, the staff has performed several specific HRA methods development activities for NPP fires, fuel cask handling, low power and shutdown conditions, and medical treatments. Participants in these programs have been aware of the SRM-directed activity to develop a single methodology, so we hope that they can all be brought together once the SRM-driven approach is completed.

At the juncture of HF and HRA, the staff has procured an NPP control room simulator and is developing a human performance test facility. These facilities will allow the staff to perform experiments rather than relying solely on collaboration in contracted facilities.

We find that the HF and HRA research program has evolved into a carefully coordinated series of projects that are extending knowledge in this area and providing useful products to regulators.

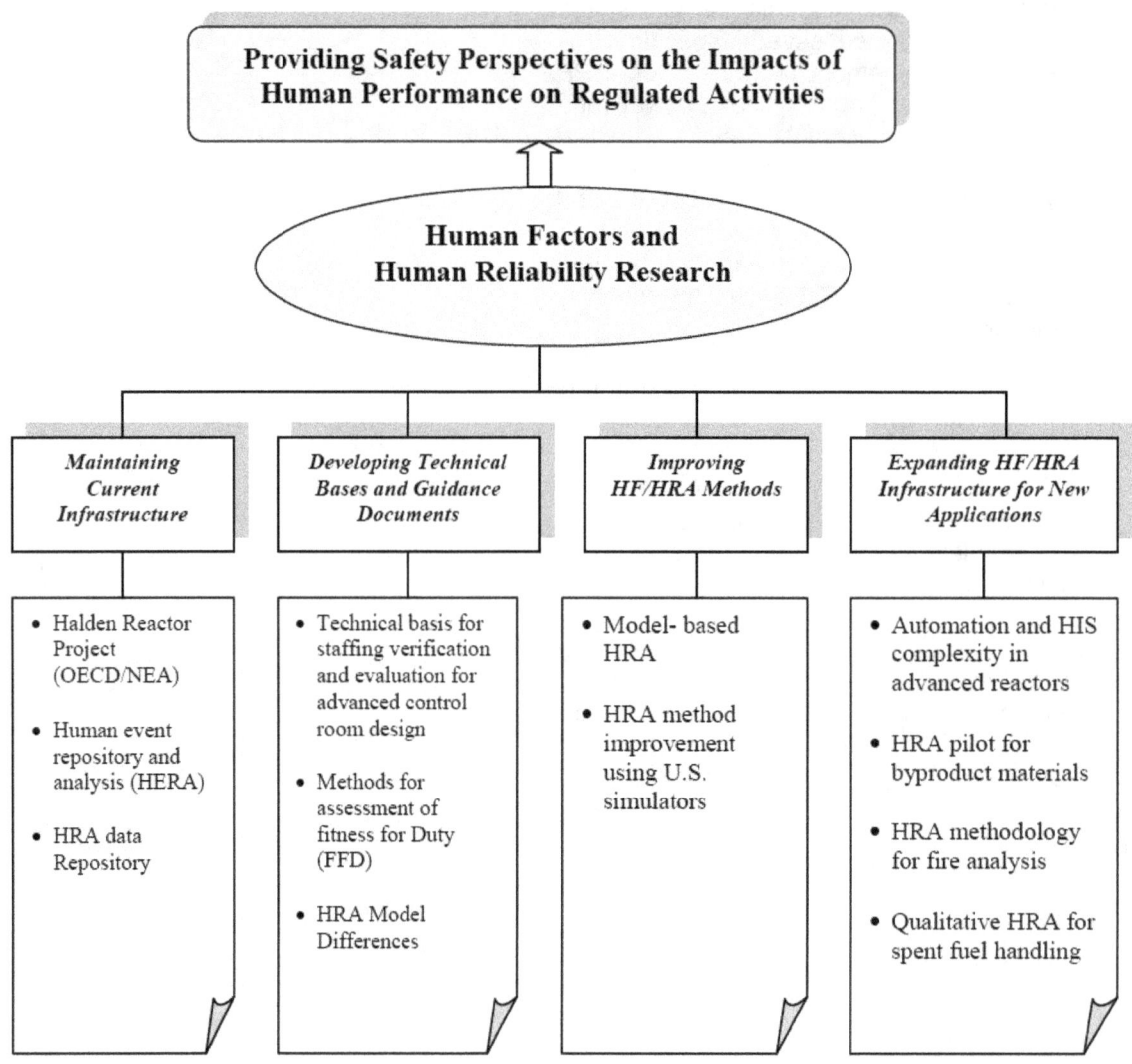

Figure 3. Current NRC Research Activities in Human Factors and Human Reliability

34

9. MATERIALS AND METALLURGY

Background

As plants age, known degradation mechanisms will continue to affect components important to safety, and new degradation mechanisms may develop. Materials and metallurgy continues to be an active area of research within the NRC. This is appropriate in light of the efforts required by the agency to address known and emerging materials degradation phenomena in aging light-water reactors (LWRs), and to monitor the effectiveness of licensees' aging management programs.

Current Research Activities

Current materials and metallurgy research activities are conducted by the Component Integrity Branch and the Corrosion and Metallurgy Branch within the Division of Engineering of RES. The former directs research focused on fracture mechanics, nondestructive-evaluation (NDE), and safety assessments and the latter on corrosion, metallurgy and advanced reactors. Topical areas of research are grouped into five broad categories:

- component integrity assessments

- nondestructive evaluation

- environmentally assisted cracking (EAC)

- proactive management of materials degradation

- steam generator tube integrity

The research activities in the five areas (see Figure 4) are appropriate and address the important materials issues. The results of these research activities will improve the agency's ability to independently evaluate licensees' efforts to prevent or mitigate

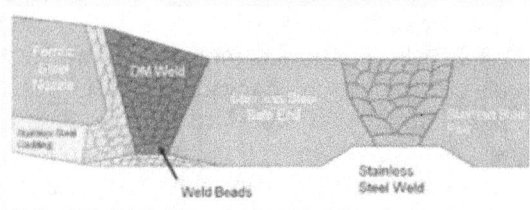

Cross-section of a dissimilar metal weld between a nozzle and safe end

environmentally assisted stress corrosion cracking and other environmental degradation mechanisms.

RES is making excellent use of domestic and international cooperative programs to accelerate progress, reduce cost, and resolve key issues related to the detection, understanding, and mitigation of materials degradation phenomena. These include the Program for the Inspection of Nickel Alloy Components (PINC and now PINC II), the Stress Corrosion Cracking and Cable Aging Program, the OECD Pipe Failure Data Exchange, the OECD Component Operational Experience, Degradation and Aging Program (CODAP), the Halden Reactor Project, the International Forum for Aging Management (IFRAM) Research and the Zorita Internals Research Project.

Component Integrity Assessments

This area of research addresses the integrity of dissimilar metal welds in piping, control rod drive mechanisms (CRDM), and pressure vessels; develops probabilistic and deterministic fracture mechanics tools, evaluates effectiveness of primary water stress-corrosion cracking (PWSCC) mitigation methods, and validates methods used to determine weld residual stresses in dissimilar metal welds. In addition to research on metallic components, work is being done to confirm the acceptability of high density polyethylene (HDPE) piping for long-term safety-related applications.

A research project is in progress to replace the previous probabilistic fracture mechanics code, PRAISE. The new code (xLPR) will incorporate more complete and realistic models for degradation, improved representations of weld residual stresses, and take advantage of modern computing power to improve fracture mechanics modeling. The new code is built on the recognition that modern piping systems typically have extremely low probability of rupture, hence xLPR. The code incorporates flaw distributions, fracture mechanics models, degradation models, and metallurgical variables in a manner similar to that used in development of the new pressurized thermal shock rule. Results from the code are being used to develop the technical basis for an RG for demonstrating compliance with the requirements of General Design Criterion 4 (GDC-4), "Environmental and Dynamic Effects Design Bases," of Appendix A, "General Design Criteria for Nuclear Power Plants," to 10 CFR Part 50. This is part of an effort to develop strategies to ensure continued compliance with GDC-4 given that many systems approved for leak before break are potentially susceptible to PWSCC (i.e., whether inspections alone or inspections plus mitigation are needed to provide assurance).

Although the initial focus of the xLPR development on primary piping integrity, the long-term vision is a modular computer code capable of addressing more generic problems associated with reactor coolant system integrity and based on verified and validated methodologies for predicting events with a low probability of failure. This is a good example of coupling a long-term goal with intermediate developments that address current regulatory needs.

Experimental programs have been initiated to obtain data on the stability characteristics of complex-shaped flaws in dissimilar metal welds and the residual stresses profiles in such welds to help validate the xLPR code.

An important input to the development of the xLPR code is the validation of estimated weld residual stresses in dissimilar metal welds. This research will include determinations of weld residual stresses in simple plates and cylinders and in complex mockups of pressurized-water reactor (PWR) components. Weld residual stresses will be measured using x-ray diffraction, neutron diffraction, and incremental deep hole drilling techniques.

HDPE materials have attractive properties for service water and buried piping applications, but they are subject to aging and degradation mechanisms that differ from those of metal piping and other types of piping conventionally used. A growing number of licensees propose to use HDPE in safety-related applications. Failure mechanisms and service life as affected by temperature, time, and flaws need to be researched. The research will include fracture-mechanics-based flaw-tolerance evaluations of base materials, as well as fusion joints.

Major progress has been made in completing and closing several research projects on reactor pressure vessel integrity and incorporating the remaining research in the component integrity assessments program. This research has led to revision of several NRC regulations, such as the pressurized thermal shock rule in 10 CFR 50.61, "Fracture Toughness Requirements for Protection Against Pressurized Thermal Shock Events," as well as related American Society of Mechanical Engineers and American Society for Testing and Materials codes and standards. Work is now in progress to revise other rules that address pressure vessel integrity such as Appendix G, "Fracture Toughness Requirements," and Appendix H, "Reactor Vessel Material Surveillance Program Requirements," to 10 CFR Part 50. A new revision of RG 1.99, "Radiation Embrittlement of Reactor Vessel Materials," is in development. The guide was last revised in 1988 and substantial changes in our understanding of the characterization of irradiation embrittlement have occurred that need to be addressed in regulatory guidance. A new RG for pressurized thermal shock also is being developed. The previous RG 1.154 has been withdrawn.

Work is continuing on the development of an understanding of and models to describe the statistical nature of fracture toughness behavior of ferritic reactor pressure vessel steels and to validate probabilistic fracture mechanics codes for vessels.

Nondestructive Evaluation

There are currently three major projects in Nondestructive Evaluation: "Evaluation of the Effectiveness & Reliability of NDE of Vessels & Piping," "Assess Emerging NDE for Dissimilar Metal Welds," and "NDE of High Density Polyethylene."

The project "Evaluation of the Effectiveness & Reliability of NDE of Vessels & Piping" is focused on quantifying the reliability of a broad range of NDE techniques used in power plant inservice inspection (ISI) programs. This task was initiated in 2007 to evaluate the accuracy and reliability of the NDE methods and to provide recommendations to the staff to improve the effectiveness and adequacy of ISI programs. This research, which is expected to be completed in 2012, covers the entire spectrum of NDE techniques used in reactor construction, inservice inspection, and repairs. This work is a core effort in the research program and presumably followon efforts will be initiated to stay abreast of developments in industry.

The project "Assess Emerging NDE for Dissimilar Metal Welds" provides an assessment of the reliability of emerging and deployed methods using signal analysis techniques and the experimental validation of crack morphology analysis for degraded weld materials. The complexity of weld microstructure and the morphology of cracks in welds represent significant challenges for NDE. HDPE piping and fittings are becoming more widely used for repair, replacement, and new plant construction, and a better understanding of the effectiveness and reliability of NDE of these components is needed.

There is also exploratory research to assess the use of in situ or continuous monitoring methods. The intent is to identify, in concert with the nuclear industry, those sensors and techniques that have the most promising commercial viability and fill a critical inspection or monitoring need.

Environmentally Assisted Cracking

EAC is a generic term for the various stress corrosion cracking (SCC) mechanisms that can be active in BWRs and PWRs. These complex phenomena are influenced by applied and residual stresses, water chemistry, radiation exposure, temperature, material composition, microstructure, and fabrication history. In BWRs, these mechanisms are known as intergranular stress corrosion cracking (IGSCC) and irradiation assisted stress corrosion cracking (IASCC). The IGSCC occurs on stainless steel welds outside the core regions and the IASCC occurs in radiation hardened stainless steel core internals. In PWRs the dominant mechanisms are PWSCC and IASCC. These phenomena occur on susceptible nickel base alloys and dissimilar metal welds. In recent years, EAC has occurred in components internal to the vessels in BWRs and PWRs, and in reactor vessel penetrations and dissimilar metal welds. In addition to primary system leakage, EAC can lead to serious secondary damage. At Davis Besse, PWSCC cracks in CRDM nozzles led to leakage of borated coolant and accelerated corrosion of the pressure vessel head.

Licensees have implemented improved inspection methods, resistant materials and repair procedures, and improved water chemistries to prevent EAC. These actions have been effective in reducing the frequency of EAC events, but they have not eliminated the problem. EAC continues to occur in nuclear power plants as metal components age and radiation exposure increases.

The NRC staff must maintain capabilities to evaluate licensees' analyses of the active degradation phenomena in their plants, and the effectiveness of implemented or proposed mitigation methods. The research projects now underway are designed to ensure that the NRC has the necessary technical understanding of the root causes of the various environmental degradation phenomena, their underlying mechanisms, and the long-term reliability of mitigation methods.

PWSCC mitigation measures include replacement with more resistant materials such as Alloy 690 and Alloys 52 and 152 weld materials. In the case of dissimilar metal welds, it is often impractical to actually remove susceptible materials like Alloy 182 and the problem is mitigated by using an overlay of resistant material. Strong research programs are in place to assess the long-term reliability of these replacement materials and the effectiveness of weld overlays in arresting cracks that have developed in a susceptible underlying material to provide adequate performance for the life of a plant, crack growth rates typically must be less than 10^{-11} meter(s) per second (m/s). Even with modern techniques, each test can take months to perform.

It is more difficult to characterize the effectiveness of mitigation measures for IASCC, because prototypical materials are difficult to obtain. The materials in the research were irradiated in the Halden Boiling Water Reactor (HBWR) or in BOR-60. The HBWR irradiations are fairly prototypical, but the fluences that can be achieved do not represent end-of-life conditions. The BOR-60 irradiations have lower fluences, but fluxes are much higher than prototypical. The NRC is participating in the Zorita Internals Research Project. This is an international collaborative effort involving the NRC, the Spanish regulator, Consejo de Seguridad Nuclear, and industry in which reactor internals materials from the Zorita reactor will be harvested and tested. This is an opportunity not only to obtain higher fluence materials irradiated under prototypical conditions, but also to obtain material for larger specimens to validate the application of data obtained with small laboratory specimens to larger components.

In addition to the Zorita project, the NRC is an active participant in other international efforts relating to materials degradation such as IFRAM Research. Such international efforts were pioneered by the NRC over 30 years ago and remain valuable and cost effective today.

Proactive Management of Materials Degradation

The nuclear industry and the NRC have often been surprised by unexpected material degradation events. Several years ago the NRC initiated a project "Proactive Material Degradation Assessment" to identify materials and systems in LWRs where degradation can reasonably be expected to occur in the future. With such knowledge, current inspection and monitoring programs at plants could be reviewed and modified as needed to provide early identification of incipient degradation.

The staff completed Phase 1 of the project in 2008. A comprehensive assessment of the likelihood and safety significance of possible environmental degradation mechanisms was completed for approximately 1900 BWR and PWR components, and NUREG/CR-6923, "Expert Panel Report on Proactive Materials

Degradation Assessment," documenting this work was issued. The objective of Phase 2 of the project was to establish agreements with industry and international organizations to define research tasks addressing the identified issues of greatest concern.

The Zorita Internals Research Program was such an activity. In cooperation with DOE, the initial assessment documented in NUREG/CR-6923 is being expanded to encompass longer lifetimes and additional materials like concrete and cable insulation in addition to the reactor vessel and core internals and piping, an Expanded Proactive Materials Degradation Assessment (EPMDA).

A program, "Analysis of Long Term Operation (LTO) Materials Degradation Issues," has been established to provide an independent analysis of the results of EPMDA, international periodic safety reviews, and operating experience to evaluate effectiveness of current aging management programs (AMPs) and support the development of enhanced AMPs in license renewal guidance documents for subsequent renewal periods.

The ACRS supported the initial vision of the Proactive Materials Degradation Assessment Program and has recommended its continuation. In NUREG-1635 Volume 9, ACRS commented that the program seems to have lost its momentum and focus. The work published in NUREG/CR-6923 identified target items of low knowledge and high-to-medium susceptibility to degradation. There was an expectation that the next step in the proactive process should be the validation of one or more of these predictions by experimental means or by focused inspections at operating plants.

However, in light of the additional operating experience highlighted through the license renewal process, the current approach of revisiting the assessment of materials degradation and expanding the scope to include concrete and cable insulation seems warranted, and will provide a firmer basis for experimental programs or new NDE or monitoring programs.

Steam Generator Tube Integrity

Rupture of steam generator tubes in PWRs can lead to accidents that allow radioactive materials released from the core to bypass the reactor containment and enter directly into the environment. Severe accidents involving containment bypass can be risk dominant in some PWRs.

The major technical areas of the NRC research program in this area include: assessment of inspection reliability, evaluation of in-service inspection technology, evaluation and experimental validation of tube integrity and integrity prediction modeling, and evaluation and experimental validation of degradation modes.

Current work includes independent assessment of the reliability of the automated eddy current analysis techniques that are now being used by industry.

Integrity studies are continuing on the use of the idealization of a rectangular crack for estimating failure pressure and leak rate for complex crack geometries.

Secondary side cracking occurs primarily in crevices where chemistry conditions can be much more severe that the nominal water chemistry. Such environments could be aggressive even for SCC resistant materials such as Alloy 690. Work that has established a better understanding of the nature of crevice behavior is being completed as part of this program.

The effectiveness of the NRC research program is again enhanced by cooperative efforts. The NRC is currently participating in the fourth, 5-year term of the International Steam Generator Tube Integrity Program (ISG-TIP-4). Current participants include organizations from Canada, France, Japan, Korea, and the United States.

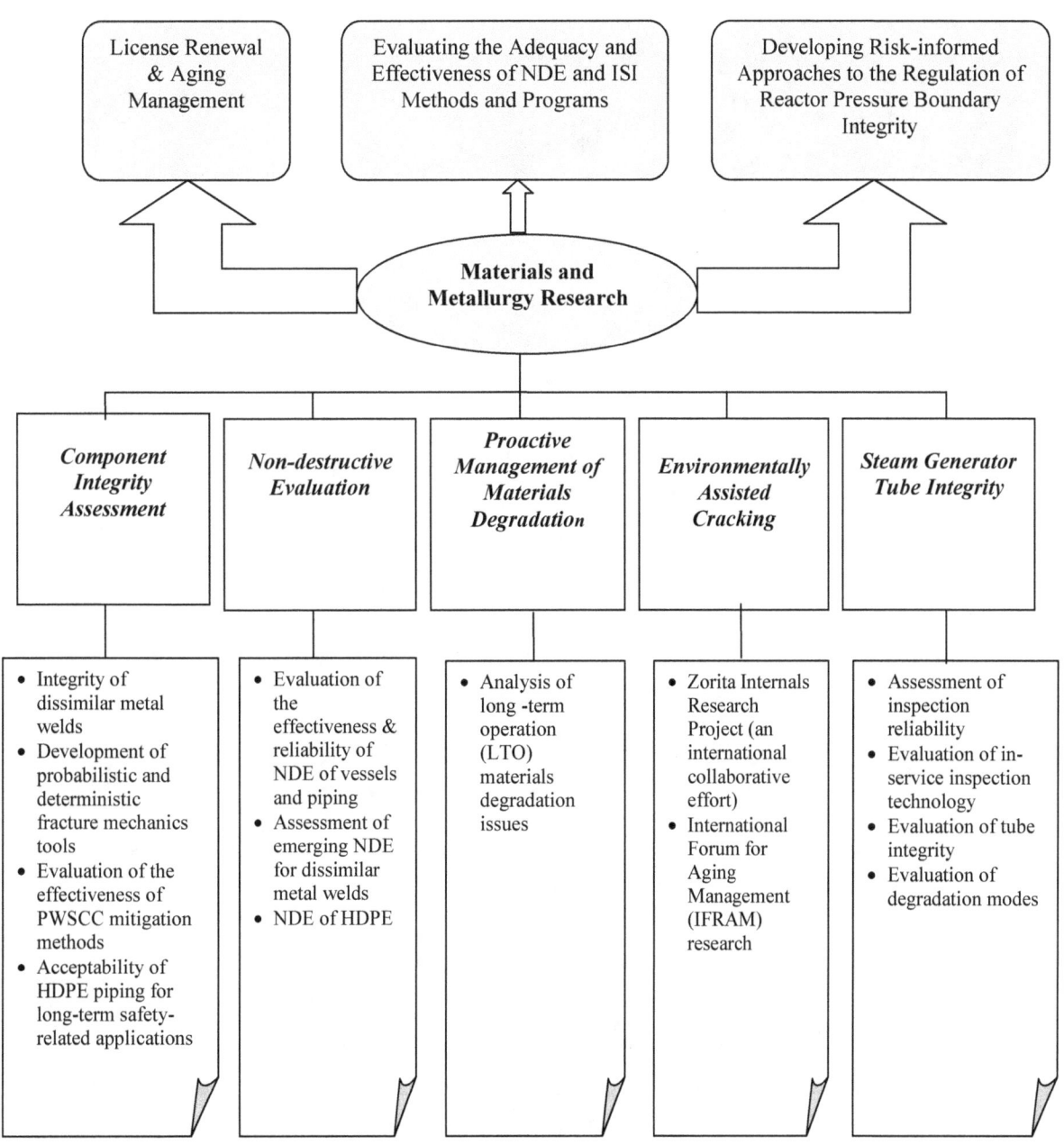

Figure 4. Current NRC Research Activities in Materials and Metallurgy

10. NEUTRONICS AND CRITICALITY SAFETY

Background

The technical capability to conduct neutronics and criticality evaluations at or very near the current state-of-the-art level is an essential core competency of the NRC. This capability is in particularly high demand in recent years. Notable areas where the capability is being used include:

- Evaluation of burnup credit for analysis of spent fuel criticality

- Licensing of the mixed oxide fuel fabrication facility located on the Savannah River site

- Certification of new LWR core designs and confirmation of applicant accident analyses

- Confirmatory analyses of BWR power uprate applications

The primary computational resources available to the staff for neutronics and criticality analysis are the SCALE suite of computer codes maintained in cooperation with the DOE at the ORNL and the PARCS reactor kinetics code that can be used in conjunction with the TRACE computer code for thermal-hydraulics analyses. These computational resources also support the MELCOR accident analysis code and the FRAPCON/FRAPTRAN suite of fuel behavior computer codes. The relationships among the computational resources, the programs they support, and the uses of code outputs are shown in the sidebar.

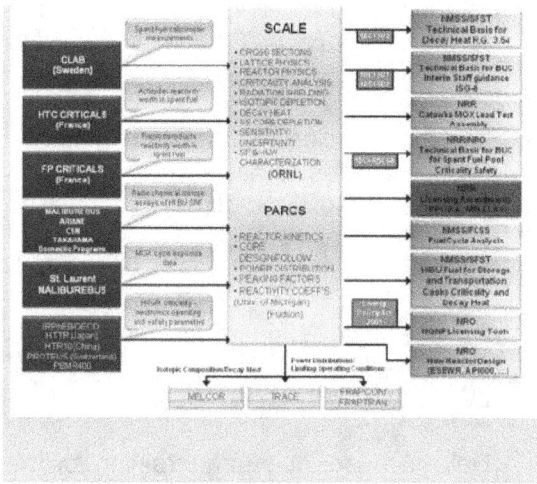

Current Research Activities

Current research activities in the area of neutronics and criticality are as follows:

- Maintain the PARCS code and train NRC staff in use and limitations of the code. This activity supports staff evaluations of the MELLA protocol for power uprates in BWRs.

- Investigate criticality issues of fuel with greater than 5 percent enrichment in ^{235}U.

- Modify SCALE architecture to enable parallel processing in analysis of accident sequences.

- Develop nuclear data libraries for NGNP designs.

- Investigate reactor physics for the prismatic and pebble-bed designs for NGNP.

- Update the SCALE code suite to enable nuclear analysis for NGNP.

- Enhance computational capabilities to enable prediction of nuclide composition, decay heat, criticality and associated uncertainties for gas reactor fuel at burnups of 80 and 160 GWd/t per metric ton.

- Enhance radiation transport methods and databases to assess shielding and evaluate irradiation degradation of the structural integrity of graphite and metal components in NGNP.

- Provide user support for the PARCS code package for use in analysis of LWR designs.

- Deliver a user guide and training for the application of SCALE/TRITON in the generation of cross section libraries for the PARCS code.

- Upgrade SCALE capabilities in the areas of in-core power peaking, reactor stability and control, core monitoring, fuel burnup, radionuclide inventories for accident source terms, safe shutdown, decay heat power, radiation sources and attenuation, and nuclear criticality safety.

- Provide support for the development and use of benchmark data for validation of burnup credit criticality analyses including the MINERVE data base from France that is part of the OECD/NEA Burnup Credit Benchmark Exercise. The work will establish the technical basis for expanded guidance on burnup credit in spent nuclear fuel storage and transport applications.

All the programs are well supported by user needs or by NRC obligations under the Energy Policy Act and appear to be progressing well.

Even greater demands on neutronics and criticality analyses can be anticipated in the future. Much of this arises because licensees of existing and planned reactors are going to ever greater lengths to optimize core designs. Optimization often involves reductions in the symmetry of the core and contributes to the need for larger scale computations. Furthermore, there is greater demand for uncertainty and sensitivity analyses of computational results. Together these evolutions are leading to more intensive uses of computational capabilities and create the need for faster code execution. The NRC is sponsoring efforts to develop parallel computational capabilities in the SCALE code suite.

Assessment and Recommendations

11. OPERATIONAL EXPERIENCE

Background

When properly documented and analyzed, operational data provide an invaluable source of information that can be used to refine the regulatory process and improve its effectiveness. With the increased use of risk information in the regulatory process, operational data become even more important to ensure the risk information has an adequate basis for use in decisionmaking. In the operational experience research area, there are several programs aimed at capturing appropriate data, analyzing the data, and providing the tools necessary to use the data in the regulatory process. In addition, the operational data is used as a measure of the regulatory effectiveness and as input to the annual report to Congress on significant operating events.

Current Research Activities

The RES Data Collection, Analysis, and Trending Programs and their relationship to each other are graphically shown in Figure 5. For example, activity "Access to INPO's EPIX System" is providing a mechanism whereby NRC staff has access to some of the Institute of Nuclear Power Operations (INPO) operational data through Equipment Performance and Information Exchange System (EPIX) so it can be used by the NRC. The data is then coded and integrated with other sources of information under the research activity "Reactor Operating Experience Data." This integrated data can then be used as inputs to PRA models, special reliability studies, and operating experience analyses under activity N6631, "Computational Support for Risk Applications." All of these ultimately provide input to programs and processes such as the Significance Determination

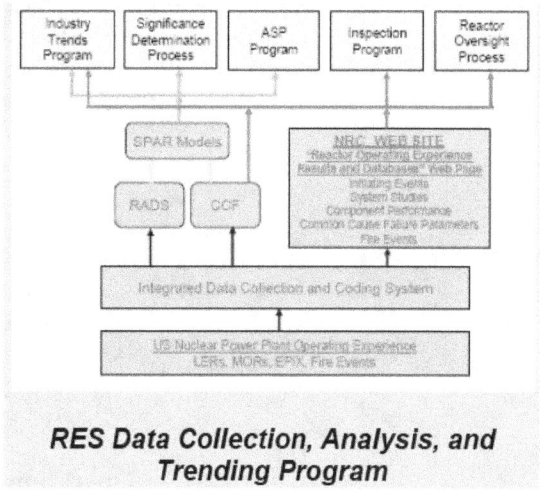

RES Data Collection, Analysis, and Trending Program

Process, Industry Trends Program, Inspection Program, and the ASP Program.

The operational experience research program is an integrated system of data collection, data coding, and tools for use of the data. Although the programs do not fall within a single branch within RES, the coordination between the various groups within RES and other program offices appears to be good and results in products that improve the regulatory process. These efforts yield improved NRC PRA models for evaluating existing plants as well as new reactor designs submitted for certification. The operational research activities are providing quantitative measures for the effectiveness of the regulatory process.

Assessment and Recommendations

The operational experience research program is being managed and executed in a manner consistent with the needs of the NRC. It provides data and tools necessary for regulatory decisionmaking and for the assessment of regulatory effectiveness. There appears to be good coordination between the research staff and the user organizations. The operational experience research program should be continued.

The events at the Fukushima Dai-ichi Nuclear Complex highlight the need to strengthen and integrate the various onsite emergency response capabilities. RES should initiate a research program to provide the necessary technical basis needed for enhancement and integration of the various onsite emergency response capabilities and the development of command and control and decisionmaking structures.

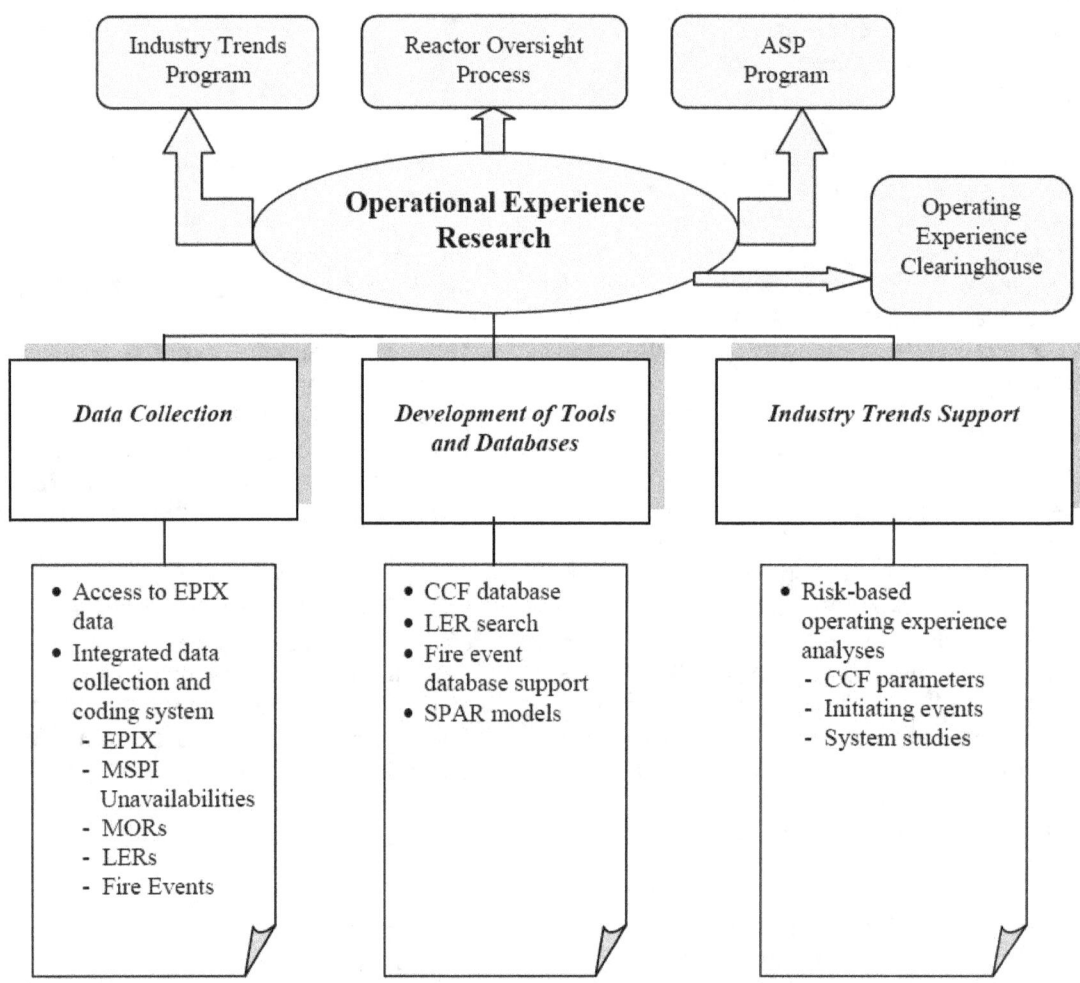

Figure 5. Current NRC Research Activities in Operational Experience

12. PROBABILISTIC RISK ASSESSMENT

Background

PRA is an essential technology for the NRC. The regulatory process for operating reactors continues to make greater use of risk information. Increasingly more comprehensive PRAs are being used to inform design decisions for new reactors and will be used in future regulatory oversight of those designs. Risk significance evaluations for emergent issues provide an important perspective to inform NRC policies, priorities, and regulatory decisions. The NRC must have state-of-the-art PRA capabilities to support all of these regulatory functions.

In its 2010 report to the Commission on the review and evaluation of the NRC Safety Research Program (NUREG-1635, Vol. 9), the ACRS noted that much of the recent research work in this area has focused on applications of existing PRA models and data to support the reactor oversight process for the current reactor fleet. Extensions of PRA scope and the development of new methods have not been priorities. That focus continues to guide much of the current PRA research and planned activities.

Current Research Activities

The Division of Risk Analysis of RES has adopted a goal-oriented framework to organize its research programs. The four fundamental program goals are:

(1) Support the Reactor Oversight and Operating Experience programs.

(2) Remove obstacles to implementation of risk-informed regulation.

(3) Expand PRA infrastructure to encompass new and advanced reactor concepts and designs.

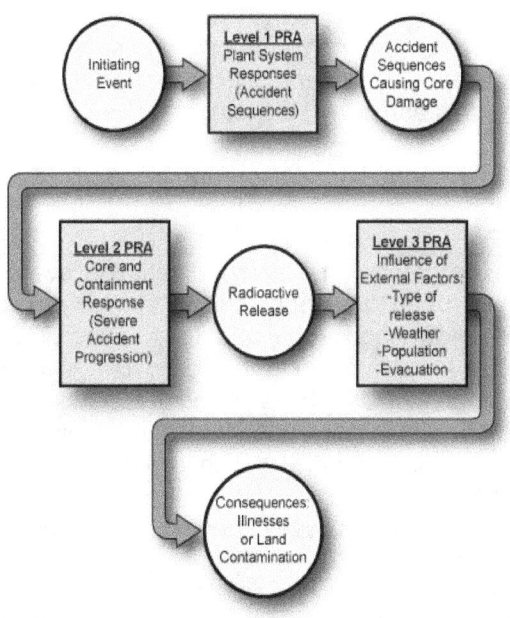

(4) Support continuous advancement in PRA state-of-the-art and state-of-practice.

In its 2010 report to the Commission, ACRS noted that this type of framework "could provide a basis for developing long-term research goals that can provide a constant vision of where the PRA research program should be headed." The committee continues to encourage that broad perspective and an emphasis on future research needs.

The Division of Risk Analysis of RES has developed close working relationships with its counterparts in NRR and NRO to define specific user needs and priorities for short-, intermediate-, and long-term research projects. Collaborative agreements and memoranda of understanding with organizations such as EPRI and NASA are being used effectively to share knowledge and resources. ACRS supports these efforts and encourages further collaboration for mutual benefits in areas of focused technical needs.

Figure 6 shows the program goals and lists the current research topics in each area. The following items briefly summarize selected research projects that extend the scope of current NRC-supported risk models and analysis methods.

<u>SPAR Models Extension and Support</u>

<u>SPAR All Hazards Models (SPAR-AHZ):</u> Limited models for external initiating events have been developed for 16 plants, based on information submitted for the individual plant examination of external events analyses. Of those models, only the Shearon Harris model has been validated by comparisons with plant-specific information and design features, including internal fire analyses that were performed for transition to NFPA-805. A project is underway to validate the other 15 SPAR-AHZ models, and create new models for other plants.

<u>Low-power/shutdown (LP/SD) models:</u> The current scope of the SPAR models includes LP/SD plant operating states for only eight plants.

<u>Digital I&C PRA</u>

Current projects are focused primarily on the representation of software failures in a pilot model of a research reactor digital protection system. Candidate quantitative software reliability models have been selected and will be evaluated to determine their capabilities for the pilot analysis. Efforts also continue on the development of methods to define coherent hardware and software failure modes for more complex integrated digital instrumentation, protection, and control systems.

<u>Examining Risk-Informed Applications for New Reactor Designs</u>

This NRO project supports the performance of tabletop exercises (using new reactor SPAR models) to examine proposed risk-informed applications for new reactor designs. Examples include risk-informed in-service inspection, surveillance test intervals, allowed outage times, SSC categorization, maintenance rule issues, and reactor oversight significance determinations. Conclusions from these exercises will support preliminary assessments of the effectiveness of currently applied risk metrics and guidance in RG 1.174, "An Approach for Using Probabilistic Risk Assessment in Risk-Informed Decisions on Plant-Specific Changes to the Licensing Basis," to maintain desired safety margins for new reactors.

<u>Level 3 PRA Project</u>

This new project involves the development of a full-scope comprehensive Level 3 PRA for an operating plant site. The planned PRA scope covers all plant operating modes, spent fuel storage pools, and onsite dry cask storage facilities. The analyses will include internal initiating events, internal hazards, and external events. Site-level risk from accidents that may involve multiple units will also be evaluated.

<u>Advanced Level 2 and Level 3 Modeling</u>

This project examines the development of advanced techniques for the integration of Level 2 and Level 3 PRA models. Methods such as dynamic PRA models and fully integrated thermal-hydraulic simulation tools will be evaluated for improved characterization of the time evolution and phenomenological aspects of accident scenario progression.

Assessment and Recommendations

The program structure in Figure 6 facilitates examination of the fundamental goals of each project and the balance among research priorities. A strong emphasis on responsiveness to immediate user needs has resulted in program priorities and resource allocations that are influenced primarily by short-term requirements to support existing PRA model infrastructure. For example, the PRA research program managers indicated that approximately 80 percent of FY 2011 PRA resources were allocated to projects in support of goal 1, approximately 15 percent were allocated to goal 2, and approximately 5 percent were allocated collectively to goals 3 and 4[2].

Maintenance and updating of SPAR models, extension of those models to include a consistent level of detail for each operating reactor, comparisons with industry PRA models, and compilation of operating experience data are important tasks to support risk-informed decisionmaking and balanced reactor oversight. However, those support activities for existing SPAR models do not typically require extension of fundamental analytical knowledge, methods, or modeling techniques beyond the current capabilities of trained PRA practitioners.

The ACRS believes that it is important to extend PRA knowledge and analysis capabilities in the regional offices by providing risk engineers and inspectors with direct "hands-on" involvement and technical responsibilities for their respective plant SPAR models. Maintenance of the models in each office would also improve implementation of model changes for consistency with plant-specific modifications and updates to operating experience data. Oversight and periodic consultation with RES headquarters experts would be needed to ensure consistency in general modeling methods and assumptions as well as for guidance during major extensions to the scope of each model. The ACRS recommends that RES should examine the feasibility for increased sharing of SPAR model maintenance and support activities

---

[2] These estimates are very approximate and do not include resources expended to support fire PRA and human reliability analysis. They are noted here only to illustrate broad relative allocations for the purposes of this report.

among the regional offices, headquarters staff, and contractors thus allowing RES to focus more effectively on advancement of state-of-the-art PRA capabilities.

The full-scope Level 3 PRA project will provide an integrated risk model context to help define the scope and requirements for intermediate- and long-term research. Some of the difficult technical issues are known and are being addressed by current projects, but at a relatively low level of effort. Other specific challenges will emerge as the Level 3 PRA models are assembled and analysts discover inevitable unexpected problems. Consistent assessment of the site-specific risks from internal hazards and external events, spent fuel accidents, and integration of the Level 1, Level 2, and Level 3 PRA models will also help inform the agency's response to issues that have been identified in the wake of the accidents at the Fukushima Dai-ichi site. ACRS recommends that RES should use evolving knowledge and insights from the Level 3 PRA project as input for risk-informing initiatives for specific research that may be proposed as a direct consequence of agency responses to the Fukushima Dai-ichi accidents.

Comprehensive PRAs for all new reactors and risk-informed applications for an increasing number of upgraded operating plants must include coherent models for complex digital instrumentation, protection, and control systems. ACRS continues to emphasize the need to understand and define unambiguous failure modes for these systems that address both hardware and software. Those failure modes are essential to provide a consistent context for analyst understanding of the PRA model elements, development of an effective logic structure, compilation and characterization of the supporting operating experience and data, and application of the most appropriate quantification techniques. ACRS recommends that failure mode research efforts should have top priority, and clear project goals should be set. Efforts to examine possible logic model structures, quantification techniques, and data should remain subsidiary until the relevant failure modes are established.

Methods to systematically identify, document, and quantify sources of uncertainty are not applied consistently throughout the agency. ACRS recommends that RES should initiate efforts to ensure that an appropriate characterization of uncertainty is performed in all agency analyses. Explicit acknowledgement, quantification, and communication of uncertainties will be a key element for rational understanding of health risk to the public and its contributors as PRA results are expanded to integrate the risks from hazards such as fires, floods, seismic events, and other extreme natural phenomena. A clear understanding of the uncertainties and their sources is also essential for risk-informed regulatory decisionmaking and rational selection among possible options to manage identified risks.

Small modular reactor designs that include several units at a single site with common external infrastructure introduce additional challenges that must be addressed to evaluate the risk from potential single-unit, multiple-unit, and site-level accidents. Analytical capabilities, modeling techniques, and tools to support a risk-informed licensing process for those reactors must be developed and tested well in advance of the first applications. ACRS recommends that RES should extend the initial efforts in this area to more fully address these site-level issues according to a schedule that adequately anticipates the first design certification applications.

The 10-year anniversary of the tragedies on September 11, 2001 has passed. Substantial efforts have been implemented to improve security against physical and cyber attacks on operating nuclear power plants in this country. New plants will further integrate security protections from the initial stages of their designs. Despite these efforts, concerns remain that the applied security controls may not be allocated optimally to cope with the full spectrum of potential threats. ACRS believes that the fundamental risk assessment framework and analysis techniques should be applied to address these concerns and to assess proposed protection strategies for future evolving threats. RES should initiate a research project and pilot applications to examine these methods.

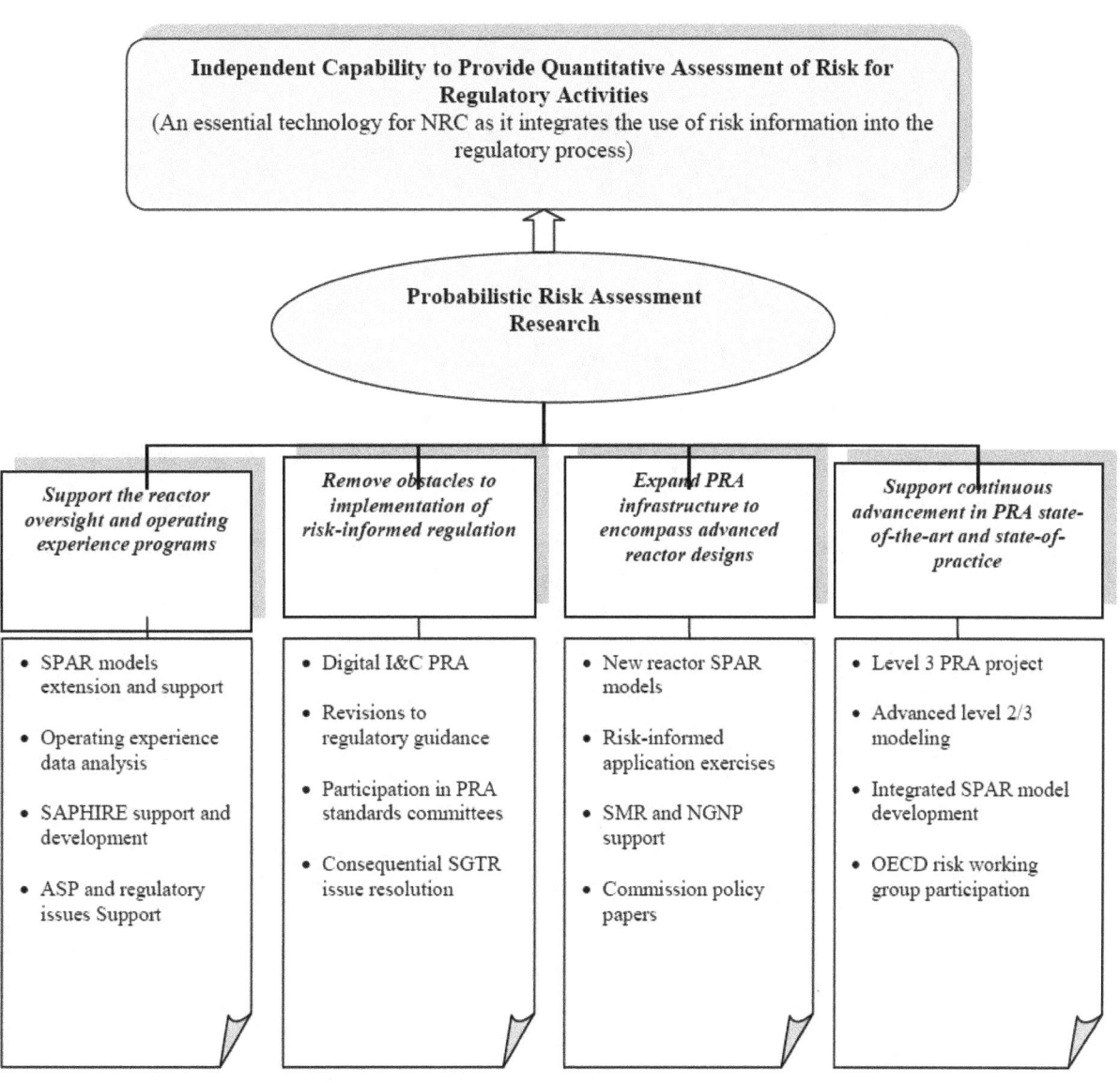

Figure 6. Current NRC Research Activities in PRA

13. RADIATION PROTECTION

Background

The NRC continues to maintain a program of research related to radiation protection in the areas of:

- risks from radiation

- the sciences of internal and external dosimetry

- the fate and transport of radioactive materials in the human body and in the environment

A major thrust of this research is to collect, analyze, and disseminate information on occupational exposures reported to the NRC by licensees. This information is used to track the effectiveness of licensees' As Low As is Reasonably Achievable programs and will form the basis for future studies to evaluate the health effects of this group of workers. Another important thrust is to develop and maintain tools for assessments related to licensing, siting, environmental performance, and the decontamination and decommissioning of licensed facilities. Additionally, the NRC continues to participate in international standards setting efforts to exchange technical information for the benefit of all participating organizations.

Current Research Activities

The current NRC research activities in the area of radiation protection are depicted in Figure 7. The research is focused on development and maintenance of health effect and dose calculation tools, emerging health effects and dosimetry research, and participation in a number of national and international collaborative radiation protection activities. There are also some efforts in the preparation of exposure and abnormal occurrence reports. These are all

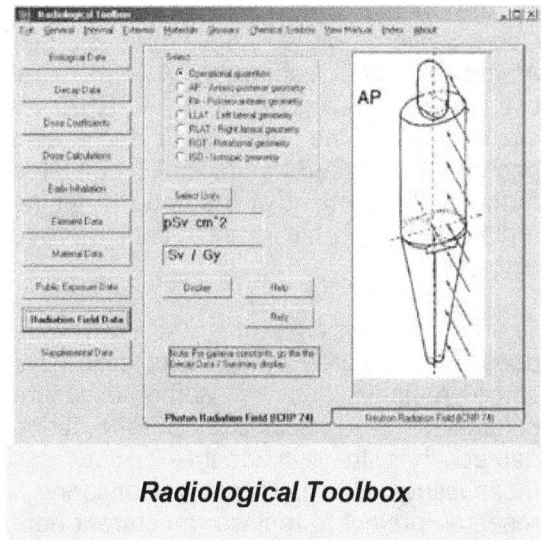

Radiological Toolbox

essential activities and need to be sustained.

Health Effects and Dose Calculation Tools

Health effects and dose calculation tools are used to model and assess the health implications of radioactive exposure and contamination.

VARSKIN: The NRC sponsored the development of the VARSKIN code in the 1980s to assist licensees in demonstrating that they have approved radiation protection programs that include established protocols for calculating and documenting the dose attributable to radioactive contamination of the skin. Since that time, the code has been significantly enhanced by adding the ability to model three-dimensional sources (cylinders, spheres, and slabs) with materials placed between the source and skin (including air gaps that attenuate the beta particles). In addition, the code incorporated a user interface that greatly simplified data entry and increased efficiency in calculating skin dose.

Since the release of VARSKIN 3 in 2004, the NRC staff has compared its dose calculations for various energies and at various skin depths, with doses calculated by the Monte Carlo N-Particle Transport Code System (MCNP) developed by Los Alamos National Laboratory (LANL). That comparison indicated that VARSKIN 3 overestimated the dose with increasing photon energy. RES has recently completed further enhancement of the code including replacement for the code's photon dose algorithm.

The current version of the VARSKIN code does not accurately predict beta dose from some radionuclides for some exposure conditions. The accuracy of the code decreases with skin depth. To address these issues, RES is currently sponsoring a research project to replace the current beta dose model to enable the code to more accurately model beta dosimetry resulting from contamination on the skin or on protective clothing covering the skin. The secondary objective of this project is to further enhance the code's functionality.

Radiological Toolbox: The NRC developed the radiological toolbox as a means to quickly access databases needed for radiation protection, shielding, and dosimetry calculations. The toolbox is essentially an electronic handbook with limited computational capabilities beyond those of unit conversion. Further revisions of the toolbox continue as the need for additional data is identified by NRC staff and other users. The toolbox contains radioactive decay data, biokinetic data, internal and external dose coefficients, elemental composition of a large number of materials, radiation interaction coefficients, kerma coefficients, and other tabular data of interest to the health physicist, radiological engineer, and others working in fields involving radiation. The toolbox includes a means to export the tabular data to an Excel worksheet for use in further calculations. It operates in a Windows environment.

RES is currently sponsoring a research project at ORNL to update the current version of the radiological toolbox. This update will include re-implementation of the software as Java based and provide additional databases and computational tools to increase the utility of the radiological toolbox for a wider range of applications. The feasibility of development of a mobile version of the radiological toolbox will also be explored.

RADTRAD 4.0: The RADionuclide Transport and Removal And Dose Estimation (RADTRAD) code uses a combination of tables and numerical models, based on simplified source term parameters, to determine the time-dependent dose at user-specified locations for a given accident scenario. It also provides the inventory, decay chain, and dose conversion factor tables needed for the dose calculation. The RADTRAD code can be used to assess occupational radiation exposures (typically in the control room) to estimate site boundary doses, and to estimate dose attenuation caused by modification of a facility or accident sequence. RADTRAD was rewritten in 2009 as a plug-in to the Symbolic Nuclear Analysis Package (SNAP). SNAP removes the need for analysts to use the text-based entry methods by providing a powerful, flexible, and easy-to-use graphical user interface (GUI).

RES is currently sponsoring a research project to add more features to the RADTRAD plug in to SNAP before the final release of RADTRAD 4.0. This project will also provide support for development of the combined RADTRAD 4.0 user manual, theory manual, and V&V documentation as a NUREG/CR report.

Participation in National and International Radiation Protection Activities

RES is actively engaged in monitoring and participating in a number of national (e.g., the National Council on Radiation

Protection and Measurements (NCRP), the National Academy of Sciences (NAS, and international committees (the International Commission on Radiological Protection, (ICRP), and Committees of the International Atomic Energy Agency, (IAEA)). NRC research also is well leveraged by working with a number of collaborating agencies, including the U.S. Environmental Protection Agency. Such activities promote consistency and coherence in regulatory applications of radiation protection and health effects research among NRC programs, as well as those of other Federal and State regulatory agencies.

Emerging Health Effects and Dosimetry Research

Impact of Reduced Dose Limits on NRC Licensed Activities: The objective of this project is to revise NUREG/CR-6112, "Impact of Reduced Dose Limits on NRC-Licensed Activities: Major Issues in the Implementation of ICRP/NCRP Dose Limit Recommendations." Additional information will be gathered about the actual dose distributions from stakeholders and information on compliance with the current and proposed limits to help the NRC/RES staff determine whether to impose a dose constraint and, if so, what value would be the most appropriate. At a minimum, this information will include an estimate of the impact of the following three occupational dose-limit options: (1) 10 rem in 5 consecutive years provided that no more than 5 rem is received in any of the years as recommended by ICRP Publication 60, or (2) simple and straightforward dose rate of 2 rem per year, or (3) conformance with the NCRP-recommended dose limit in its Report 116, "Limitation of Exposure to Ionizing Radiation," that allows a maximum of 5 rem per year less the accumulated dose in rem (i.e., tens of mSv) reaching to the individual worker's age in years, or keep the current status.

Analysis of Cancer Incidence and Mortality in Population Living Near NRC-Licensed

Nuclear Facilities: Efforts are underway to have the U.S. NAS conducted a study analyzing the cancer risk of populations living near NRC-licensed facilities, including power reactors and fuel cycle facilities (e.g., fuel enrichment and fabrication plants). This study will update and expand on the 1990 U.S. National Cancer Institute report, "Cancer in Populations Living Near Nuclear Facilities." The staff plans to use this report as a scientifically defensible resource to aid in addressing continued stakeholder concerns about perceived elevated cancer rates in populations near reactors, including cancer incidence (i.e., being diagnosed with cancer but not necessarily dying from the disease).

Assessment and Recommendations

The ACRS believes that the staff has developed an appropriate and robust research program in the areas of radiation protection. This program includes radiation protection of workers and radiological assessments related to radiation exposure and health risks around NRC-licensed nuclear facilities.

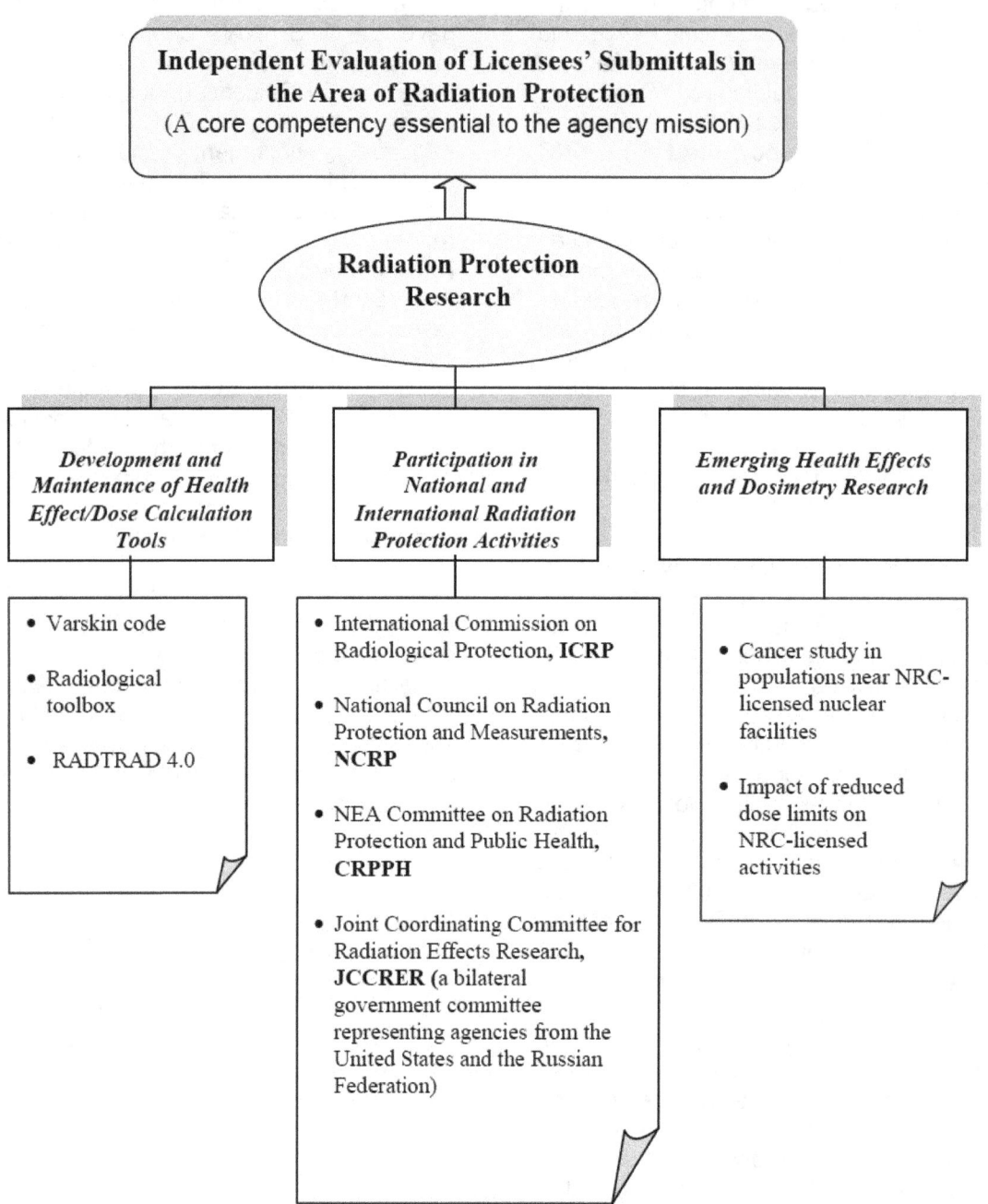

Figure 7. Current NRC Research Activities in Radiation Protection

14. NUCLEAR MATERIALS AND WASTE

Background

The NRC continues to maintain a robust program of research related to nuclear materials and radioactive waste topics related to licensing, facility siting, facility environmental performance, and the decontamination and decommissioning of licensed facilities.

Current Research Activities

The staff is engaged in a number of research and development projects. These projects cover a wide range of topics from fundamental studies on the fate and transport of radioactive materials in the environment to the predictability of material science questions related to spent fuel pools and dry storage casks. These activities will support both specific licensing actions and generic issues. Four projects are summarized below.

Application of Uncertainty Methods to Nuclear Reactor Sites

This project is intended to streamline the Pacific Northwest National Laboratory uncertainty methodology for practical implementation during environmental and early site permit (ESP) application reviews. This project will also develop tools for accessing and interpreting retrievable databases (e.g., United States Geological Survey (USGS), U.S. Environmental Protection Agency (EPA), and State databases) for use in predicting performance through numerical modeling (e.g., GMS, MODFLOW, and MT3D) and assessing uncertainties of the hydrologic conceptual models, parameters and scenarios. NRO licensing staff and PNNL scientists and engineers are reviewing ESP applications for the next generation of nuclear power plants. Using site-specific

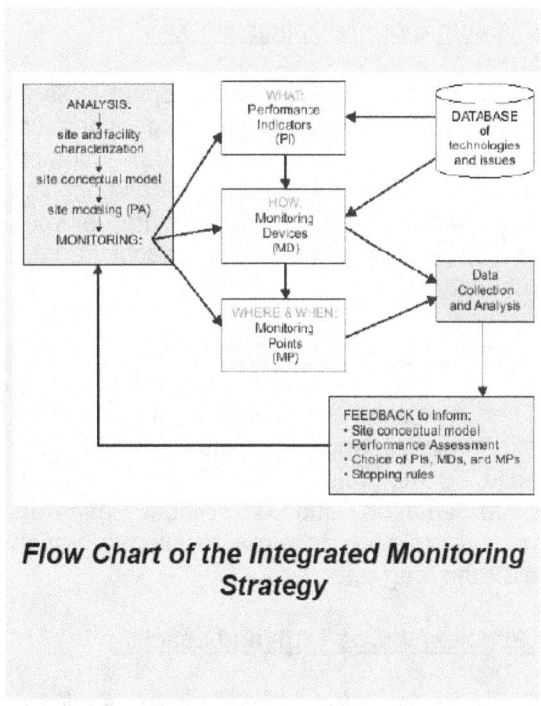

Flow Chart of the Integrated Monitoring Strategy

information from the ESP applications, they will draft environmental impact statements for these proposed new plants.

Many important technical review issues surround the assessment of site-specific ground-water and surface-water conditions and behaviors that influence radionuclide transport. Under a variety of reasonable release scenarios, radionuclide transport through the ground-water pathway is a key component in assessing potential exposures to the public. NRC regional inspectors and RES scientists are evaluating new ground-water monitoring programs developed by licensees in response to the Nuclear Energy Institute initiative on ground-water protection. Action plans for operating nuclear power plants in response to this initiative include identification of conceptual hydrogeologic models and the reliability of methods to detect and monitor liquid releases. Decisions on the need for and choice of remediation technologies depend on these ground-water monitoring data and their attendant uncertainties.

Uranium Bioremediation, Phase I

This project is aimed at developing tools to model short- and long-term biogeochemical processes controlling uranium mobility during and after bioremediation of 1) shallow groundwater uranium plumes and 2) remediation of in situ leach uranium recovery sites.

Uranium Solid Phase Behavior

This project is intended to develop experimental and modeling details of the solid minerals developed during bioremediation and how those minerals behave relative to uranium sequestration over the long term.

Complex Source Term and Leaching

The objective of this project is to develop a mechanistically sound and practical approach to describe releases from materials with complex source terms, such as slags and concrete.

15. SEISMIC AND STRUCTURAL ENGINEERING

Background

Nuclear power plants are designed to cope with substantial seismic events—up to at least the safe shutdown earthquake. Earthquakes of greater magnitude have the potential to damage the power plant and barriers to radionuclide release. Earthquake damage can be a common cause mechanism for the failure of systems to prevent accidents and systems to mitigate accidents– extending even to the final line of defense in depth of public evacuation and emergency planning.

The seismicity of the Western United States is relatively well known. The focus of current attention is on the seismicity of the country east of the Rocky Mountains where events are less common and most of the nation's nuclear power plants are located. There are seismic centers in the Central and Eastern United States that have the potential to produce large earthquakes. The New Madrid, Charleston, Northeast, and Eastern Tennessee seismic centers are examples. Events at these seismic centers can affect large areas since seismic attenuation is thought to be less in the Central and Eastern United States than in the Western U.S. Recently, the USGS has revised expected "return frequencies" for large earthquakes at some of these seismic centers to values higher than used for seismic hazard analysis in the past. The NRC has revived its programs of research into seismic events and the responses of nuclear power plants to such events.

Current Research Activities

The primary focus of the NRC seismic and structural engineering research program is supporting regulatory activities, especially in the areas of certification of new plant designs and combined license applications.

Computed maximum tsunami wave amplitude in the Atlantic Basin generated by a Mw 8.8 earthquake in the Caribbean source zone

The current research activities can be broadly categorized as:

- seismic hazard characterization

- structural engineering with emphasis on seismic and impact loading

- international collaborations in seismic research

- development of regulatory guides

Within the broad research category of "seismic hazard characterization," research is being conducted to:

- develop an up-to-date seismic source characterization for the Central and Eastern United States that includes a full assessment and incorporation of uncertainties

- develop practical procedures for implementation of the guidelines for updating probabilistic seismic hazard assessments (PSHAs)—the so-called "SSHAC guidelines" for the use of expert opinion in analysis of seismic issues

- develop improved models of seismic attenuation in the Central and Eastern United States

The result of the source characterization study will be a consensus regional seismic source database to be used in a PSHA to determine the seismic hazard at any particular site. The study is being conducted using the SSHAC guidelines. It is a cooperative effort of the NRC, DOE, and EPRI.

The staff also is performing independent assessments of the seismic hazard for selected sites in the Central and Eastern United States and performing sensitivity analyses of critical input parameters. This will help resolve discrepancies between NRC staff and industry.

Based on the work done by the staff in collaboration with the USGS to characterize tsunami threats, the staff is undertaking a technology and knowledge transfer for improved tsunami hazard assessment methods and modeling tools and exploring tsunami resistant design technologies and the development of tsunami PRA methods.

In the area of earthquake engineering, research is being performed to provide the technical basis for new regulatory guides on seismic isolation technologies and the impact of multi-dimensional loading and incoherent ground motion in soil-structure-interaction analyses. This research will also provide the technical basis for an update to RG 1.208, "A Performance-Based Approach to Define

the Site-Specific Earthquake Ground Motion," which currently states that applicants should consider inclined waves in certain geologic conditions, but provides no guidance as to how this should be done.

Structural engineering research is currently focused on the effects of aging degradation on the capacity of containment structures and the benchmarking of numerical simulations of aircraft impact on nuclear reactor facilities to provide improved insights on modeling and damage criteria, and increase confidence on these analyses. The work on containments will develop methods and results that the NRC can use to assess the condition, strength and operability, of containments when significant degradation has occurred.

The NRC staff is continuing its collaboration with Japan on the large-scale testing of SSCs during shaking that simulates earthquakes. Further collaborations are taking place in understanding the impacts of the Kashiwazaki-Kariwa earthquakes. Data from this earthquake presents a unique opportunity to benchmark codes and analysis methods.

Assessment and Recommendations

The NRC seismic and structural research program in support of regulatory activities has been exemplary. There is a well-developed research plan that has been broadly reviewed for both technical quality and programmatic impact.

16. SEVERE ACCIDENTS AND SOURCE TERM

Background

The NRC makes use of its severe accident expertise and analysis capabilities to support regulatory decisions for operating nuclear power plants and for certifying new and advanced reactor designs. Severe accident analysis tools also help the staff in its transition to a more risk-informed regulatory framework. In addition, as evidenced from the Fukushima events, severe accident analysis tools and insights are essential for coping with such events by identifying potential accident progression scenarios and radiological releases.

The RES long-term severe accident and containment response evaluation development plan focuses on two areas: maintenance and development of the MELCOR and TEXAS computer codes and continued collaboration in international experimental research programs.

Considerable investment was made by the NRC in the past to achieve the current levels of severe accident understanding. Recent NRC investments have been limited to a level that allows continuation of required analysis and risk-informed activities including:

- Code model enhancements, such as targeted assessment of MELCOR for containment-related phenomena (e.g., DBA peak pressure and related containment analyses).

- Code model development and assessment for non-LWR (e.g., high-temperature gas-cooled reactor) applications.

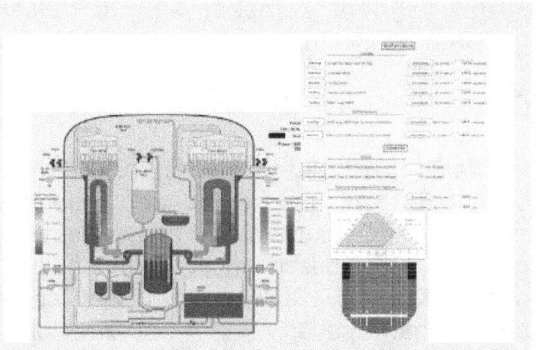

MELCOR Accident Simulation Using the Symbolic Nuclear Analysis Package (MASS)

The RES continues developing the MASS (MELCOR Accident Simulation using the Symbolic Nuclear Analysis Package) for simplifying MELCOR input deck development and result evaluation. MASS provides a convenient display system for the MELCOR code users to define an accident sequence and to observe the plant response. The MASS interface will facilitate the use of MELCOR by the NRC staff.

- Design certification efforts of new reactor designs (e.g., evolutionary power reactor (EPR), advanced boiling-water reactor (ABWR), economic simplified boiling-water reactor (ESBWR), AP1000, and advanced pressurized-water reactor (APWR).

- Risk-informing severe accident management strategy development, severe accident induced steam generator tube rupture evaluations, and SOARCA support.

- Code maintenance and user support activities, such as user-interface enhancements (e.g., SNAP capability and the MACCS interface module) and international technical review meetings including Cooperative Severe Accident Research Program (CSARP), MELCOR Cooperative Assessment Program (MCAP), and European MELCOR User Group (EMUG).

Current Research Activities

Figure 8 illustrates the current NRC severe-accident and source-term research activities. All severe-accident research activities are focused on development and usage of the MELCOR or the TEXAS code and participation in international collaborative severe accident research programs. There are also some efforts underway on severe accident knowledge management.

The NRC leverages its resources by relying on collaborations in international research programs in the Pacific Rim (Japan and Korea) as well as Europe (France, Switzerland, and Germany). The body of knowledge gained from the NRC's past experimental work and those obtained from international experimental programs are systematically incorporated into the MELCOR accident analysis code.

Organized by the NRC, CSARP is an annual forum for exchanging severe accident research findings. Twenty foreign countries currently participate in CSARP. One significant outcome of this effort is the adoption of the MELCOR code by other countries and institutions as an analytical tool for severe accident analyses.

MELCOR Code Development and Usage

The MELCOR code is a fully integrated, engineering-level computer code whose primary purpose is to model accident progression in current nuclear power plants, new and advanced reactor designs, and some nonreactor systems such as spent fuel pools. In 1982, the original MELCOR code development was focused on supporting Level 2 PRAs. Thus, code development included the reactor core, coolant system, and safety systems as well as containment systems in a less detailed manner than more mechanistic thermal-hydraulic and fuel rod codes. Efforts continue to consolidate the physical models and capabilities of more detailed severe accident codes, such as CONTAIN, into MELCOR. This consolidation effort is designed to ultimately provide an efficient state-of-the-art code for severe accident analyses.

The MELCOR code and its atmospheric dispersion and radiological consequence calculations model (MACCS) have been extensively used for the SOARCA project. Several recent modifications enhanced MELCOR capabilities for the SOARCA effort. These include (1) new default parameters, either as input records or sensitivity coefficients, that best capture experimental observations and improve code numerical robustness, (2) addition of a model to simulate the thermo-mechanical collapse of fuel rods with highly oxidized cladding at high temperatures, and (3) enhancement of the SPARC pool scrubbing model to treat fission product vapors.

In addition, as a part of the SOARCA project, NUREG/CR-7008, "MELCOR Best Modeling Practices," is being developed. This report describes how MELCOR modeling capabilities were used to represent important, but uncertain, aspects of severe accident behavior. This description includes choices made among alternative modeling options offered through code input, changes to selected input parameters from those offered as "default" values, and in some cases, application of user-generated "models" to represent features of plant response to a severe accident that are not directly available in MELCOR.

To support NRC design certification efforts, MELCOR analyses were performed for several new reactor designs (e.g., EPR,

ABWR, ESBWR, APWR, and AP1000). In addition to these new LWR designs, the NRC has decided to use MELCOR to provide confirmatory analysis of small modular reactors, including the HTGR, and the iPWRs that will be submitted to the NRC for licensing review. Several model development efforts are required to simulate unique features of these reactor designs. In the case of DOE-funded NGNP HTGR, model development efforts are nearing completion that will allow MELCOR to simulate the release and transport of fission products within the fuel particles and graphite matrix as well as the generation and transport of dust and fission product aerosols within various HTGR systems and components. NGNP MELCOR development efforts are performed in collaboration with the DOE-funded NGNP program. An MELCOR model was developed for one iPWR for the NuScale plant, and some preliminary severe accident simulations were completed.

To support iPWR design certification submittals, the ACRS notes that additional model development, and in some cases, experimental validation studies, may be needed. The ACRS also notes that the applicability of an LWR severe accident code to gas-cooled reactor systems may require extensive validation efforts (if DOE continues to fund deployment of the NGNP).

Various workshops and user training programs are conducted each year to support MELCOR users. For example, the MELCOR Code Assessment Program (MCAP) is an annual technical review meeting that focuses on MELCOR code development and assessment and provides a forum for users to present and discuss their experiences. In addition, several European organizations have initiated EMUG meetings.

The ACRS supports the NRC's work-in-kind participation in the European Severe Accident Research NETwork of excellence (SARNET). Such efforts ensure that the

NRC is cognizant of analysis approaches taken by other international agencies and contributes to multi-national convergence of evaluation methodologies and the success of the multi-national design evaluation program.

Other Code Activities

NRC severe accident research activities primarily focus on MELCOR. However, there are other codes NRC maintains for specific severe accident analyses, including:

- **TEXAS** - A stand-alone code that models fuel coolant interaction (FCI) phenomena in the vessel or containment. MELCOR analysis provides initial and boundary conditions (pressure, temperature, melts mass and composition) for TEXAS.

- **MACCS** - A computer code for calculating atmospheric dispersion and offsite radiological consequences of fission product releases to the environment. MACCS requires data on the magnitude and composition of radioactive materials released to the environment and the associated energy content, time, release elevation, and duration of release.

- **CONTAIN** - A specialized analysis tool to perform containment response analysis in LWR plants under postulated DBA and beyond-DBA events to predict thermal-hydraulic conditions inside containment.

During the ESBWR design certification review process, it was recognized that the containment and reactor coolant system are closely coupled during the long-term cooling phase of the accident. As noted before, efforts are underway to consolidate the physical models and capabilities of CONTAIN into MELCOR. ACRS concurs with the staff plans to ultimately rely on MELCOR for containment analyses.

Collaborative Severe Accident Experimental Research Programs

The NRC participates in several collaborative experimental severe accident research programs, which continue to provide key data for MELCOR model development and assessment. Table 1 summarizes the objectives and scope of these international collaborative severe accident research programs. NRC participation in an international program is evaluated using established criteria to ensure that the expected value of such collaboration to the NRC is well worth the investment.

ACRS continues to support the NRC approach to leverage resources by obtaining data through participation in international experimental collaborations. ACRS also notes that models in the NRC's system-level code, MELCOR, were developed from a database that is primarily focused on in-vessel testing with significantly more PWR-specific experiments. Hence, it is anticipated that BWR-specific and ex-vessel modeling enhancements will be identified as more information from the recent events in Fukushima Dai-ichi nuclear complex in Japan becomes available. ACRS recommends that the NRC expand its current severe accident research program to obtain the required data to enhance and validate models that are found to be deficient.

Severe Accident Research Knowledge Management

As a part of its commitment to the agency effort in knowledge management (KM), RES has initiated efforts to collect and catalogue experimental data and models used to describe severe accident phenomena for submittal to the NRC's Agencywide Documents Access & Management System (ADAMS).

The NRC invested heavily in severe accident research through major experimental and model development studies to achieve the current understanding of the progression and the radiological consequences of severe accidents. The ACRS strongly endorses the RES KM efforts to preserve the information and insights gained from this research. The ACRS also encourages RES to seek collaborations with EPRI and SARNET2 for preserving and consolidating experimental data.

Assessment and Recommendations

ACRS supports the strategy that the NRC staff has developed to support regulatory decisions for severe accidents via computer code development validated by experimental data analysis and evaluation. This approach has successfully allowed the NRC to maintain and update its modeling capabilities for severe accident analyses. Planned program extensions and continuations of these collaborations are well worth the investment. ACRS notes that ongoing MELCOR Fukushima Dai-ichi assessments may identify some deficiencies in BWR-specific and ex-vessel modeling capabilities and recommends that if such deficiencies are identified, the NRC expand their current severe accident research program to participate in efforts to obtain the required data to enhance and validate models that are found to be deficient.

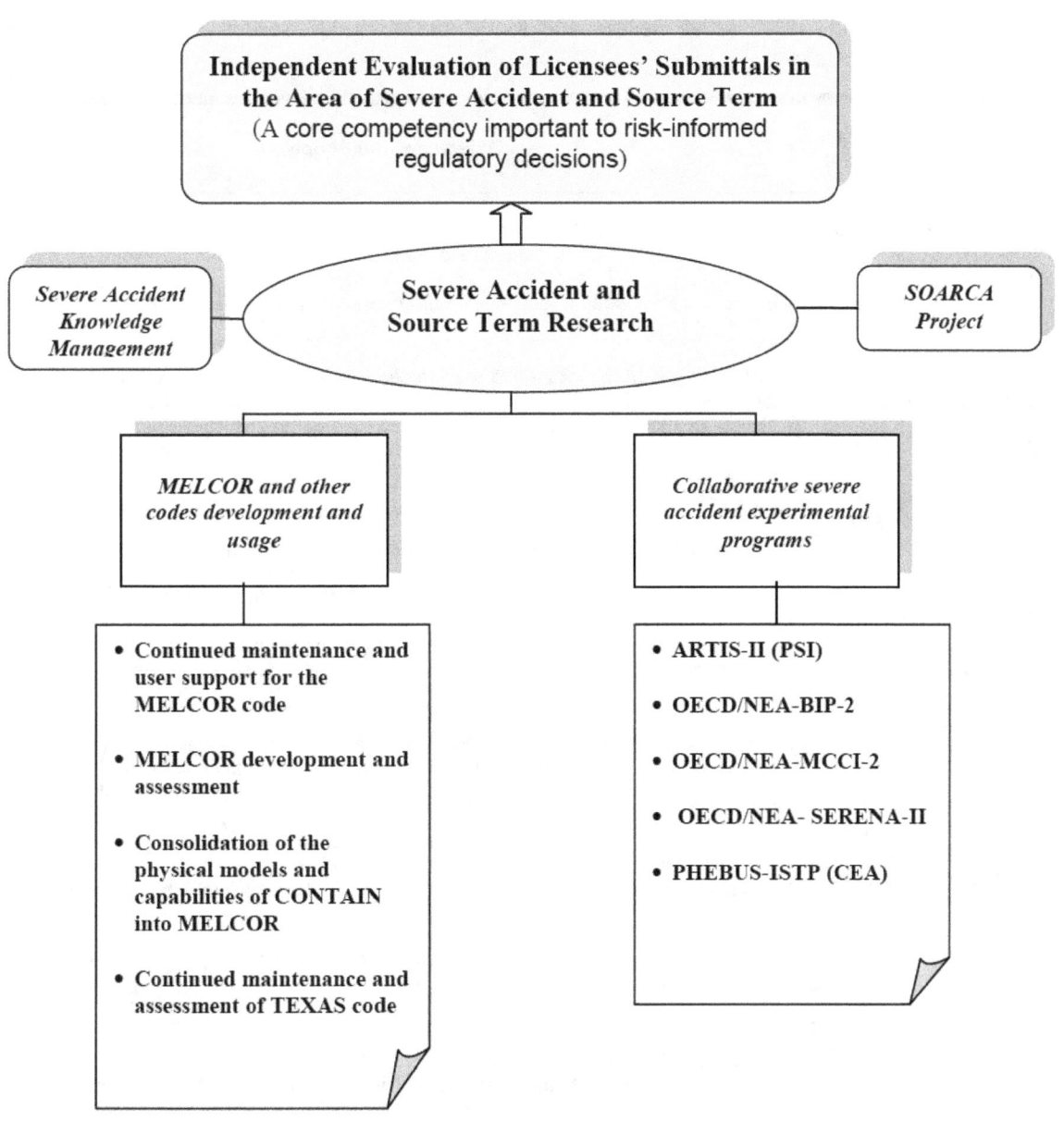

Figure 8. Current NRC Research Activities in Severe Accident and Source Term

Table 1 International Collaborative Severe Accident Experimental Research Programs with NRC Participation

Program/Performing Organization (Country)	Objectives and Scope
ARTIST (Aerosol Trapping In a Steam Generator), Paul Scherrer Institute (Switzerland)	The original tests, ARTIST-I, (completed) focused on measuring aerosol decontamination in the secondary side of steam generators during steam generator tube rupture (SGTR) accidents. ARTIST-II experiments examine issues of particle bounce, breakup, and re-suspension discovered during the ARTIST-I tests.
OECD/NEA BIP (Behavior of Iodine Project), AECL (Canada)	BIP (completed) was initiated to provide separate effect tests and modeling studies of iodine behavior in a containment following a severe accident. The NRC recently signed an agreement to participate in the BIP-2 program, which will perform more detailed evaluations of iodine absorption on painted containment surfaces and organic iodine formation.
OECD/NEA - SERENA (Steam Explosion Resolution for Nuclear Applications) **KROTOS** Experiments, CEA (France) **TROI** Experiments, KAERI (Korea)	The SERENA program was established to assess the capabilities of fuel-coolant interaction (FCI) computer codes to predict steam explosion-induced loads in reactor situations. Phase II of the SERENA project is using the complementary features of the TROI (KAERI) and KROTOS (CEA) corium facilities. Experimental research is supplemented with analytical activities so as to improve the FCI codes and strengthen confidence in their applicability to severe accident scenarios.
OECD/NEA- MCCI (Melt Coolability and Concrete Interaction) Project, ANL (United States)	The focus of the original project, MCCI-I (completed), was to investigate the coolability of molten core materials, interacting with the containment structural concrete, by an overlaying water layer. The second phase of project (MCCI-II), carried out from 2006–2010, helped bridge data gaps not fully covered in previous tests. MCCI-II also addressed the effectiveness of design features in new LWR designs for augmenting coolability, e.g., CCI tests for examining the melt behavior with underlying cooled refractory basemat, similar to the core retention concept in the EPR. The NRC is participating in one joint EdF-IRSN/CEA-NRC test to confirm early water ingress phenomena. Data will be used in a benchmark effort by participants.
PHEBUS -FP and **ISTP** CEA (France)	The PHEBUS-FP Program (completed in 2006) consisted of a series of in-pile integral experiments of: fuel degradation; fission product release; radionuclide transport through a model of reactor coolant system; and aerosol behavior in model containment. The NRC has also participated in the follow-on program, PHEBUS-ISTP. This is a collection of separate effects projects to pursue specific aspects of PHEBUS-FP findings including chemistry of iodine in the containment (EPICUR), fission product chemistry in the reactor coolant system (CHIP), iodine interactions in containment (PARIS), air oxidation of fuel cladding (MOZART), interaction of steel and cladding with boron carbide in steam (BECARRE), and fission product release from MOX and high burnup fuel in steam and air (VERDON). With the exception of VERDON, all PHEBUS-ISTP experimental projects have been completed.
OECD/NEA Sandia Fuel Project (SFP) (United States)	The goal of this project is to provide experimental data relevant for hydraulic and ignition phenomena of prototypic water reactor spent fuel assemblies. The proposed experiments will focus on thermal-hydraulic and ignition phenomena in PWR 17x17 assemblies and will supplement earlier results obtained for BWR assemblies.

17. THERMAL HYDRAULICS

Background

Evaluation of the effects of thermal hydraulic phenomena on nuclear safety has always been a central element in the conduct of the NRC's regulatory mission. Of particular importance has been, and continues to be, the capability to independently confirm thermal-hydraulic analyses in licensees' submittals.

Early thermal-hydraulic analyses of nuclear power plants employed very conservative bounding assumptions, assuring large safety margins with regard to allowable temperatures and pressures over a wide range of accident and operating conditions. With time, the experimental database and the confidence in analytical predictions have grown, allowing the NRC to consider submittals from licensees employing "best estimate", thermal-hydraulic analyses together with estimates of uncertainties. These analyses have grown ever more sophisticated. It is necessary for the NRC to continue development of state-of-the-art thermal-hydraulic computational tools and more sophisticated understanding of important thermal-hydraulic phenomena. To this end, the NRC maintains competence in the thermal-hydraulics field and capability to conduct confirmatory analysis through its research program.

Current Research Activities

The current NRC research activities in thermal hydraulics are depicted in Figure 10. The research activities focus on the development of the TRACE computer code for confirmatory analyses of a wide range of safety-significant thermal-hydraulic phenomena and supporting experiments. There is also a modest effort to develop capability in multidimensional computational fluid dynamics (CFD).

ATLAS (Advanced Thermal-Hydraulic Test Loop for Accident Simulation) Test Facility in Korea

The NRC is participating in the International Standard Problem 50 (ISP-50). The ATLAS facility will be used to simulate a direct vessel injection (DVI) line break, which is important in several new PWR designs such as APWR and AP1000. The ISP-50 participants have performed both "blind" and open calculations for the test using various safety analysis codes including ATHLET, CATHARE, MARS, RELAP5/MOD3, as well as TRACE. This will provide the participants with the opportunity to assess existing safety analysis codes against the data. The effort is coordinated by the Korea Atomic Energy Research Institute (KAERI).

TRACE Computer Code Development and Validation

In the mid-1990s, a decision was made that the several primary reactor system thermal-hydraulic codes that were in use at that time be consolidated into a single code. The several codes included RELAP5 (for LOCA), TRAC-P (for PWR LOCA), TRAC-B (for BWR LOCA), and RAMONA (for BWR stability).

The models, correlations, and solution methodologies in these codes did not reflect the state-of-the-art and required in-depth modernization. It was also recognized that they had been designed at a time when computer capabilities were limited and included many structural features, such as memory management, that were no longer needed, and increased the cost of continued code maintenance and development. The availability of graphical user interfaces and their wide acceptance also suggested the desirability of incorporating similar capability into the NRC codes. All these considerations led to extensive code consolidation, model improvements, and implementation efforts, culminating in the development and validation of the TRACE computer code.

TRACE is intended to serve as the main tool for the confirmatory analyses of a broad range of thermal-hydraulic problems for current and new reactor designs. It has the potential to offer significantly enhanced capabilities for state-of-the-art analyses of thermal-hydraulic issues. Several important technical issues, such as core stability and Anticipated Transient without Scram behavior, involve coupling between neutronics and thermal hydraulics and require that TRACE be properly coupled to a neutronics code like PARCS, an activity that has recently been completed. The integration, validation, and assessment of the TRACE/PARCS coupled code is currently under way so that it can be reliably used for confirmatory analyses. TRACE also has the capability to interface with the CONTAIN code for containment response analysis as well as with other computational tools, including MATLAB.

In response to ACRS recommendations, the staff commissioned and completed a detailed peer review of TRACE. The main objective of the TRACE code peer review was to identify the strengths and deficiencies of the code, and provide recommendations for code changes and improvements. The peer review found no major deficiencies that would introduce significant errors or preclude the use of TRACE for analysis of postulated LOCAs in operating LWRs. Certain improvements were recommended for treatment of the momentum equations, which could be particularly important for passively cooled systems, though TRACE was not explicitly reviewed for applicability to such systems. Another improvement suggested was the incorporation of additional fluid fields such as for drops and bubbles to improve predictive capabilities.

Thermal-hydraulic system codes, including TRACE, solve an intertwined structure of approximate conservation equations and empirical correlations. The uncertainties and biases inevitably introduced by such empirical procedures need to be properly addressed. Because of these uncertainties, predictions of such codes are adequately accurate only within certain ranges of parameters. The codes cannot be given blanket approval for all situations to which they might be applied. In practice, a code, such as TRACE, must be qualified by assessment against a range of data that cover the phenomena that dominate the prediction of figures of merit, such as peak clad temperature, important to the regulatory process. These dominating phenomena change with the reactor systems and accident conditions being considered. In view of this, thermal-hydraulic codes need to be assessed for analyses of a specific accident in a particular system. Because of the high uncertainties that can exist in a calculation, modifications are being made to TRACE and SNAP so that the code uncertainty can be quantified. The modifications are intended to be general, and enable the staff to evaluate uncertainty methods proposed by applicants as well as allow the staff to statistically determine the uncertainty in calculations made with TRACE and TRACE/PARCS.

RES has initiated a systematic assessment of the applicability of TRACE to analyze new reactor designs. Work on a detailed assessment of the applicability of TRACE to analyze ESBWR LOCAs, focusing on the collapsed liquid level in the reactor pressure vessel as the primary figure of merit, has already been completed. Ongoing work on assessment of the applicability of TRACE for confirmatory analyses of safety-significant thermal-hydraulic phenomena in the AP1000, the APWR, and EPR designs have also been completed to validate the use of TRACE in the design certification process. Work has been initiated to develop TRACE for applicability to iPWR designs such as NuScale and mPower.

RES has also been participating in two cooperative international agreements to obtain high-quality experimental data to refine best-estimate calculations. The first cooperative agreement, which was recently concluded, was the NUPEC BWR full-size fine-mesh bundle tests (BFBT) benchmark. Participation in this activity allowed the NRC to obtain a database of sub-channel void fractions, pressure drops, and critical power measurements from a representative BWR fuel assembly from the Nuclear Power Engineering Corporation (NUPEC) in Japan. The extensive database and contributions of the benchmark participants will be used to improve the predictive capabilities of TRACE.

A second international cooperative agreement, centered on a PWR database, has recently completed its third and final workshop. The NUPEC PWR subchannel and bundle test (PSBT) benchmark uses a high-quality database of PWR void fraction and DNB data provided by METI/JNES. This database is used to assess TRACE code models. Like the BFBT benchmark, the PSBT agreement involved the participation of international experts who took part in three workshops intended to facilitate open discussion of the database and the modeling techniques used in other thermal-hydraulic codes. Under this agreement, METI/JNES provided the experimental data, OECD/NEA is in charge of the logistics of the event, and NRC/PSU NEP prepared the benchmark specifications and coordinated the workshops.

Multi-Application Small Light Water Reactor (MASLWR) is a system-level test facility constructed by Oregon State University (OSU) to examine thermal hydraulic phenomena of importance to integral type reactors. MASLWR is the predecessor to the current NuScale integral reactor with 1:3 length-scales, 1:254 volume scale, and a 1:1 time scale and is operated under full pressure and temperature conditions of the NuScale design. Currently, NuScale is working with OSU to update this integral test facility to the latest NuScale module design in preparation for design certification testing. Through participation in the IAEA International Collaborative Standard Problem, a TRACE model for MASLWR will be developed. Predictions of this model will be compared for several steady state and transient reactor states. Test results can also be used to validate new analysis models or components for the TRACE code. The research effort aims towards the future support of licensing reviews of the NuScale reactor.

Experimental Studies of Thermal-Hydraulic Phenomena

Thermal-hydraulic phenomena involved in normal and accident conditions for LWRs are complex, and often involve the difficult-to-model flow of two-phase mixtures (steam and water). Predictions from computer codes of such phenomena need extensive experimental validation, and there are many effects, such as those involving multidimensional flows in complex geometries, where large-scale experiments are the primary means of confirming the validity of these predictions. In view of this, NRC-RES has maintained two complex experimental facilities:

- Purdue University Multidimensional Integral Test Assembly (PUMA) facility at Purdue University for BWR-related problems

- Rod Bundle Heat Transfer (RBHT) facility at Penn State University for PWR emergency core cooling problems

The PUMA is a medium-size reduced-height scaled facility and has been used in the past to perform integral LOCA tests of interest for the ESBWR design. Tests are being conducted at the PUMA facility to obtain experimental data on the void fraction distribution and fluid dynamics of a BWR suppression pool during the blowdown period. The results of these tests will be used to support the technical assessment of Generic Safety Issue 193, "BWR ECCS Suction Concerns."

In addition, an exploratory research program to develop so-called "closure relationships" for the evolution of interfacial area in two-phase flows is being undertaken at Purdue University. It is expected that when the data encompass the range of flow regimes expected in two-phase flows, then a model of interfacial area evolution will be incorporated into the TRACE code, potentially improving its accuracy and reliability. Whether such developments should be prioritized over development of the four field model recommended by the TRACE peer review group remains to be assessed. In any way, results from this Purdue program have been slow in coming and the strategy for utilizing them in TRACE still remains to be elucidated.

The RBHT facility at Penn State University was developed to address issues related to emergency core cooling, including the development of a better understanding of reflood and rewetting in realistic, bundled geometries. Currently, this facility is being used to conduct oscillating reflood tests to determine the effect of the inlet flow rate, magnitude, and frequency on peak clad temperature. There is also an option to perform steam cooling with droplet injection tests to determine the effect of dry spacer grid on droplet size and distribution. Data from RBHT tests are being used to develop improved models for reflood thermal hydraulics and the effects of spacer grids, which has been seen as significant.

In parallel, the NRC is collaborating with international groups in undertaking experiments in facilities abroad as noted below:

OECD/NEA ROSA-2: The NRC is participating in the OECD/NEA ROSA-2 Project to use the Large Scale Test Facility (LSTF) of ROSA (Rig-of-Safety Assessment) Program of (Japan Atomic Energy Agency) for studying the integral response of the core and steam generator. The full-height ROSA/LSTF integral test facility, with 1:48 volumetric scaling, is designed to investigate thermal hydraulic phenomena of interest to PWRs. The ROSA-2 Project is in progress, and is providing both integral and separate-effects thermal-hydraulic data on intermediate break LOCA and on the recovery from Steam Generator Tube Rupture events. These data are being used to assess predictive capability of thermal-hydraulic analysis codes including TRACE.

OECD/NEA-PKL-2: The OECD/NEA-PKL2 Program is designed to address thermal-hydraulic safety issues for current PWR and new PWR designs. The PKL facility is a full-height, 1:145 power and volume scaled replica of a 4-loop, 1300 MW PWR. The experiments will focus on steam generator heat transfer under shutdown conditions (e.g., loss of residual heat removal during mid-loop operations), fast cooldown transients (such as main steam line breaks), accident situations for new PWR designs (e.g., EPR Emergency Operating Procedures), and boron precipitation. The data obtained from these experiments will be used for TRACE assessment, to complement design certification reviews, where possible, and in the case of boron precipitation, to validate licensee post-LOCA long-term cooling strategies.

ISP-50, ATLAS 50 Percent DVI Line Break: ATLAS is a 1:2-height and length, 1:288-volume, and a full pressure integral test facility in Korea. The ATLAS facility will be used to simulate a DVI line break, which is important in several new PWR designs such as APWR and AP1000. Participants in the ISP-50 have performed both "blind" and open calculations for the test using various safety analysis codes including ATHLET, CATHARE, MARS, RELAP5/MOD3, as well as TRACE. Considering the very limited integral test data on DVI line breaks, this will provide the participants with the opportunity to understand the relevant thermal-hydraulic behavior and to assess existing safety analysis codes against the data. The effort will be coordinated by the KAERI.

Development of Multidimensional CFD Capabilities

The NRC currently has a modest, but high impact, effort in the area of CFD using commercial CFD codes from ANSYS Inc. (FLUENT) and CD-Adapco (STAR-CCM+). These codes provide detailed three-dimensional single-phase fluid flow information not available from system code thermal-hydraulic simulations. The CFD predictions have been useful for enhanced understanding of certain local phenomena and have played a role in resolution of a number of broad technical issues, such as induced steam generator tube failures, distribution of injected boron in the ESBWR, and spent fuel pool analyses.

Assessment and Recommendations

The staff is to be commended for the progress that has been made in developing and moving forward with incorporation of TRACE into the regulatory process. Much work remains to be done to enable its reliable use for the analysis of the new LWR designs, an urgent matter, which should be conducted with high priority. The priorities for further development of TRACE require careful evaluation.

The international collaborative efforts are also to be commended, as they take advantage of facilities that are of a scale and capability that do not currently exist in the United States Furthermore, they draw on the expertise of international partners, who have continued to maintain a very high level of capability in the thermal-hydraulics field. However, complementary development of national facilities to address safety-related thermal-hydraulics issues should be seriously considered. Such facilities would enable the retention of U.S. expertise and provide the capability to conduct experiments for supporting confirmatory thermal-hydraulic analyses of new reactor designs.

The NRC currently has in place modest, but high impact, efforts in the area of CFD through the use of commercial CFD codes. The CFD predictions have been very useful and play a role in improving the technical bases for licensing decisions.
It is inevitable that the licensees will increasingly capitalize on the extraordinary advances in computing power and computational science to resolve problems which the current generations of

thermal-hydraulic codes such as TRACE are unable to do. NRC thermal-hydraulic research has to position the agency to address such potential developments coming from the nuclear industry. Several possibilities for developing such capabilities should be evaluated, taking budgetary constraints into account. First amongst these would be participation and cost-sharing in programs with international partners to develop next-generation multidimensional CFD simulation tools aiming at a high level of transparency, V&V. Second, perhaps in conjunction with the first, consideration should be given to the formation of an NRC-US university consortium to develop such capability, capitalizing on the high level of CFD expertise that now exists in several universities. Third, and perhaps again in conjunction with some subset of the other options, the possibility of building on one of the excellent open-source CFD platforms should be considered.

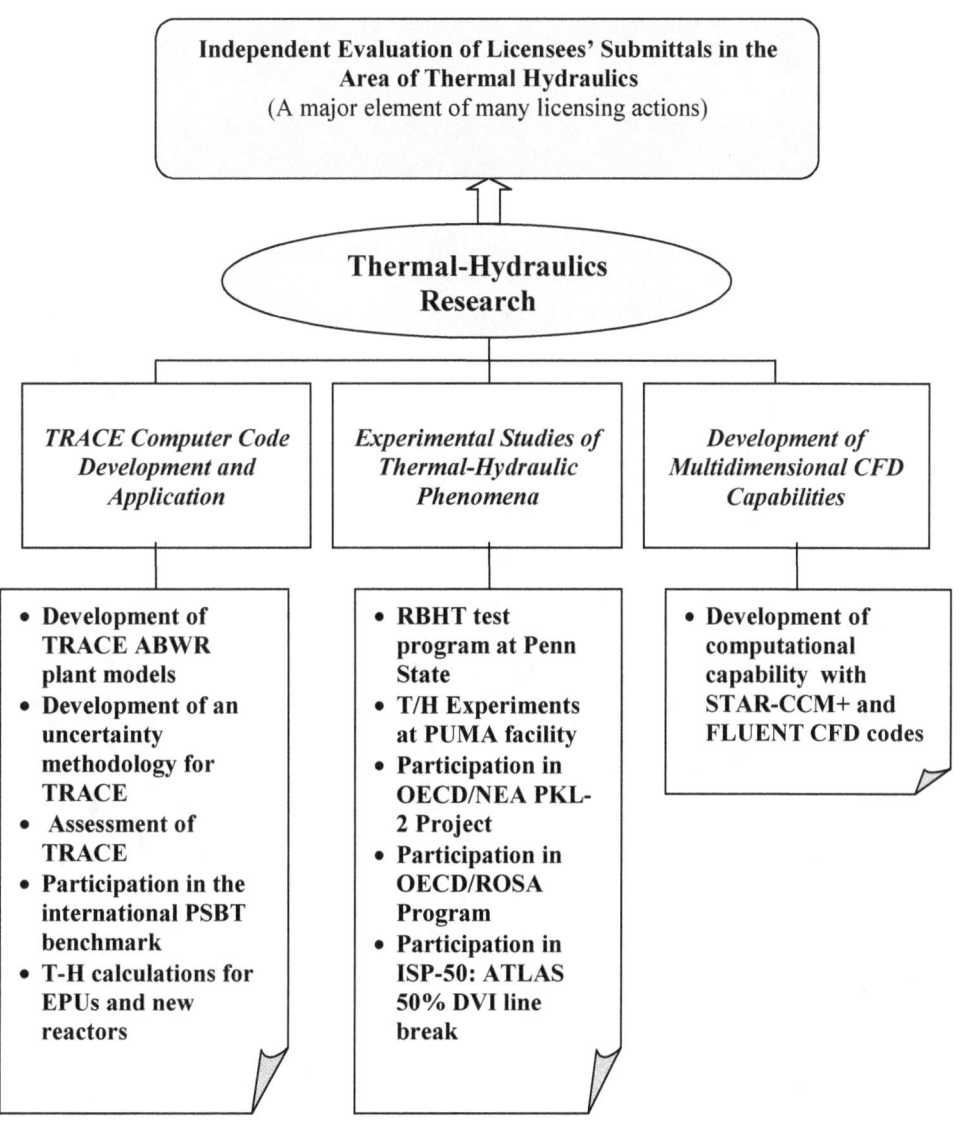

Figure 9. Current NRC Research Activities in Thermal-Hydraulics Research

18. REFERENCES

1. U.S. Nuclear Regulatory Commission, "Review and Evaluation of the Nuclear Regulatory Commission Safety Research Program," ACRS, NUREG-1635, Vol. 9, June 2010.

2. U.S. Nuclear Regulatory Commission, "Recommendations for Enhancing Reactor Safety in the 21st Century, The Near-Term Task Force Review of Insights from the Fukushima Dai-ichi Accident," July 12, 2011.

3. Letter Dated October 13, 2011, From Said Abdel-Khalik, Chairman, ACRS, to Gregory B. Jaczko, Chairman, NRC, Subject: Initial ACRS Review of: (1) the NRC Near-Term Task Force Report on Fukushima and (2) Staff's Recommended Actions to be Taken Without Delay.

4. Letter Dated November 8, 2011, From Said Abdel-Khalik, Chairman, ACRS, To Gregory B. Jaczko, Chairman, NRC, Subject: ACRS Review of Staff's Prioritization of Recommended Actions to be Taken In Response To Fukushima Lessons learned (SECY-11-0137).

5. U.S. Nuclear Regulatory Commission, "Expert Panel Report on Proactive Materials Degradation Assessment," NUREG/CR-6923, BNL-NUREG-77111-2006, February 2007.

6. *Code of Federal Regulations*, Title 10, Part 50, Section 46, Acceptance Criteria for Emergency Core Cooling Systems for Light-Water Nuclear Power Reactors.

7. U.S. Nuclear Regulatory Commission, "Severe Accident Risks: An Assessment for Five U.S. Nuclear Power Plants," NUREG-1150, December 1990.

8. Public Law 109-58, "Energy Policy Act of 2005," enacted August 8, 2005.

9. *Code of Federal Regulations*, Title 10, Part 50, Section 65, Requirements for Monitoring the Effectiveness of Maintenance at Nuclear Power Plants.

10. *Code of Federal Regulations*, Title 10, Part 54, Requirements for Renewal of Operating Licenses for Nuclear Power Plants.

11. U.S. Nuclear Regulatory Commission, "Verification and Validation of Selected Fire Models for Nuclear Power Plant Applications," NUREG-1824, Vols. 1-7, EPRI TR1011999, May 2007.

12. U.S. Nuclear Regulatory Commission, "A Phenomena Identification and Ranking Table (PIRT) Exercise for Nuclear Power Plant Fire Modeling Applications," NUREG/CR-6978, November 2008.

13. U.S. Nuclear Regulatory Commission, "Fire PRA Methodology for Nuclear Power Facilities," NUREG/CR-6850, Vols. 1-2, EPRI TR 1011989, September 2005.

14. U.S. Nuclear Regulatory Commission, "Cable Response to Live Fire (CAROLFIRE)," NUREG/CR-6931, Vols. 1-3, April 2008.

15. U.S. Nuclear Regulatory Commission, "A Short History of Fire Safety Research Sponsored by the U.S. Nuclear Regulatory Commission, 1975-2008," NUREG/BR-0364, June 2009.

16. U.S. Nuclear Regulatory Commission, "Human Factors Considerations with Respect to Emerging Technology in Nuclear Power Plants," NUREG/CR-6947, October 2008.

17. U.S. Nuclear Regulatory Commission, "Expert Panel Report on Proactive Materials Degradation Assessment," NUREG/CR-6923, February 2007.

18. U.S. Nuclear Regulatory Commission, "Guidance on the Treatment of Uncertainties Associated with PRAs in Risk-Informed Decision Making," NUREG-1855, Draft Report for Comment, November 2007.

19. U.S. Nuclear Regulatory Commission, "Estimating Loss-of-Coolant Accident (LOCA) Frequencies Through the Elicitation Process," NUREG-1829, April 2008.

20. U.S. Nuclear Regulatory Commission, "Recommendations for Probabilistic Seismic Hazard Analysis: Guidance on Uncertainty and use of Experts," NUREG/CR-6372, Volumes 1 and 2, April 1997.

21. U.S. Nuclear Regulatory Commission, Regulatory Guide 1.183, "Alternative Radiological Source Terms for Evaluating Design-Basis Accidents at Nuclear Power Reactors," July 2000.

22. U.S. Nuclear Regulatory Commission, Regulatory Guide 1.174," An Approach for Using Probabilistic Risk Assessment in Risk-Informed Decisions on Plant-Specific Changes to the Licensing Basis," Revision 1, November 2002.

23. Letter Dated September 24, 2008, From William J. Shack, Chairman, ACRS, to Dale E. Klein, Chairman, NRC, Subject: Development of the TRACE Thermal-Hydraulic System Analysis Code

NRC FORM 335 (12-2010) NRCMD 3.7	U.S. NUCLEAR REGULATORY COMMISSION **BIBLIOGRAPHIC DATA SHEET** *(See instructions on the reverse)*	1. REPORT NUMBER (Assigned by NRC, Add Vol., Supp., Rev., and Addendum Numbers, if any.) NUREG-1635, Vol. 10

2. TITLE AND SUBTITLE Review and Evaluation of the Nuclear Regulatory Commission Safety Research Program	3. DATE REPORT PUBLISHED	
	MONTH	YEAR
	October	2012
	4. FIN OR GRANT NUMBER	

5. AUTHOR(S)	6. TYPE OF REPORT
	Technical
	7. PERIOD COVERED (Inclusive Dates)

8. PERFORMING ORGANIZATION - NAME AND ADDRESS (If NRC, provide Division, Office or Region, U. S. Nuclear Regulatory Commission, and mailing address; if contractor, provide name and mailing address.)

Advisory Committee on Reactor Safeguards
U.S. Nuclear Regulatory Commission
Washington, D.C. 20555-001

9. SPONSORING ORGANIZATION - NAME AND ADDRESS (If NRC, type "Same as above", if contractor, provide NRC Division, Office or Region, U. S. Nuclear Regulatory Commission, and mailing address.)

Same as above.

10. SUPPLEMENTARY NOTES

11. ABSTRACT (200 words or less)

This report to the U.S. Nuclear Regulatory Commission (NRC) presents the observations and recommendations of the Advisory Committee on Reactor Safeguards (ACRS) concerning the NRC Safety Research Program being carried out by the Office of Nuclear Regulatory Research (RES). In its evaluation of the NRC research activities, the ACRS considered the programmatic justification for the research as well as the technical approaches and progress of the work. The evaluation identifies research crucial to the NRC mission. The interdisciplinary effect of the State-of-the-Art Reactor Consequence Analyses (SOARCA) Project is not addressed in this report. Early in the project, the ACRS provided reports on the technical approach of this activity.

12. KEY WORDS/DESCRIPTORS (List words or phrases that will assist researchers in locating the report.)		13. AVAILABILITY STATEMENT
NRC Safety Research Program	Probabilistic Risk Assessment	unlimited
Fukushima	Radiation Protection	14. SECURITY CLASSIFICATION
Advanced Non-LWR Designs	Nuclear Materiasl and Waste	(This Page)
Digital Instrumentation and Control Systems	Seismic and Structural Engineering	unclassified
Fire Safety	Severe Accidents and Source Team	(This Report)
Reactor Fuel	Thermal Hydraulics	unclassified
Human Factors and Human Reliability		15. NUMBER OF PAGES
Materials and Metallurgy		
Neutronics and Criticality Safety		16. PRICE

UNITED STATES
NUCLEAR REGULATORY COMMISSION
WASHINGTON, DC 20555-0001

OFFICIAL BUSINESS

NUREG-1635, Vol.10

Review and Evaluation of the Nuclear Regulatory
Commission Safety Research Program

October 2012